Science for Non-Specialists: The College Years

Committee on the Federal Role in
College Science Education of Non-Specialists

Commission on Human Resources

National Research Council

NATIONAL ACADEMY PRESS
Washington, D.C. 1982

The material in this report is based upon work supported by the National Science Foundation under Grant
No. SED-7912299. Any opinions, findings, and conclusions or recommendations expressed in this publication
are those of the authors and do not necessarily reflect
the views of the National Science Foundation.

Library of Congress Catalog Card Number 81-86397

International Standard Book Number 0-309-03231-8

Available from

NATIONAL ACADEMY PRESS
2101 Constitution Avenue, N.W.
Washington, D.C. 20418

Printed in the United States of America

NATIONAL RESEARCH COUNCIL
COMMISSION ON HUMAN RESOURCES

2101 Constitution Avenue   Washington, D. C. 20418

OFFICE OF THE
EXECUTIVE DIRECTOR

November 5, 1981

Dr. John Slaughter
Director
National Science Foundation
Washington, D. C. 20550

Dear Dr. Slaughter:

I am pleased to transmit with this letter the report "Science For Non-Specialists: The College Years," which is the result of the deliberations of the National Research Council's Committee on a Study of the Federal Role in College Science Education of Non-Specialists, chaired by Richard Gray of the Indiana University School of Journalism. The study was supported by the National Science Foundation under Contract SED 7912299.

This report concerns a vital area of education in U.S. colleges which has not received the emphasis it deserves. Although curriculum committees have long worried about appropriate breadth and balance in the humanities and social sciences for those studying science or engineering, little attention has been paid to the converse case. All too frequently, college graduates in the non-science areas leading to professional work in law, business, journalism, and so on have had little or no contact with science. We believe this is a serious problem that deserves early and continued attention by U.S. educators and those who support their efforts.

The National Research Council is the principal operating agency of the National Academy of Sciences and the National Academy of Engineering to serve government and other organizations

iii

The report is a readable account of the issues involved and contains a number of suggestions or possible paths for future action. The most important of these for your attention is the stress on a well-organized, aggressive role for the National Science Foundation in constructing a program which will be an agent for change in this area.

I know that there are many demands and growing restrictions upon the limited funds at your disposal. Even so, this Committee believes that appropriate science education for the non-science leaders of tomorrow--the shapers of our laws, the conveyors of our news, the managers of our enterprises--should be an urgent priority on your agenda.

Please note that the report does not say that the federal government should do it all. In fact, it assigns the major responsibility to the colleges and universities themselves. But they will need help from many sources, and the catalytic role of the federal government can be all-important.

The Commission on Human Resources is pleased to forward this report to your attention. If there is any other way we can help, please let us know.

Sincerely,

Harrison Shull
Chairman

COMMITTEE FOR A STUDY OF THE FEDERAL ROLE
IN THE COLLEGE SCIENCE EDUCATION OF NON-SPECIALISTS

ACKNOWLEDGEMENTS

In collecting information for this report, the Committee
has benefited from the support and advice of many people
and organizations.

Financial support provided for this study by the National Science Foundation is acknowledged with thanks.
Alphonse Buccino, director of the Office of Program Integration in the Science and Engineering Education Directorate, and Joel Aronson, NSF project officer, met
with the Committee throughout the year and provided
helpful information.

The Committee would especially like to acknowledge
the able efforts of Pamela Ebert-Flattau, its study director, and Gregory L. Crosby, research associate.
Linda S. Dix deserves special praise for her work in
preparing the final manuscript. Susan M. Coonrod and
Janie B. Marshall provided excellent administrative,
technical, and clerical support under considerable time
constraints.

Within the National Research Council, the Committee
received valuable counsel and assistance during all
phases of the study from Harrison Shull, chairman of the
Commission on Human Resources, and William Kelly, its
executive director. Commission members William F. Miller, John A. Moore, and Jack E. Myers reviewed the manuscript of the report and provided helpful suggestions.
Kathleen Drennan assisted in the coordination of the
preparation of the final manuscript.

The Committee is grateful for the assistance of
Philip Ritterbush, director of the Institute for Cultural Progress, who served as its consultant earlier in
the year. His summary report on the role of science in
general education at the college level was especially
helpful.

We also wish to thank the many individuals--teachers, students, administrators, members of professions, and others--who took part in the Committee's meetings, conferences, and workshops and provided much valuable information.

To all of these persons, the Committee expresses its warmest thanks.

CONTENTS

The report of a committee of the National Research Council usually is both a technical statement and a social
document. This one is no exception. It is technical in
that its findings are based on an analytical process and
are directed to a significant problem in science or
technology. It is social because it has been written by
a group of people who bring to their work diverse back-
grounds, interests, and perspectives. Also of special
social importance is the problem the present committee
was asked to address: the proper education of non-specialists who need more than a casual acquaintance with
science or engineering to discharge their professional
and civic responsibilities well in the closing years of
the twentieth century and the opening years of the
twenty-first.

The Committee for a Study of the Federal Role in College Science Education of Non-Specialists was given a
three-fold charge by its parent Commission on Human Resources: (1) to determine how science is being presented to undergraduate students who are not studying to
become scientists; (2) to recommend improvements that
may be needed in what is generally perceived to be a
neglected branch of undergraduate education; and (3) to
determine if there is a role for the federal government
in assisting colleges and universities to meet their
responsibilities to provide this important subgroup of
their students with an appropriate science education.
Early in 1980, facing a mid-1981 deadline that was all
too near, the Committee set to work with the assistance
of a small staff to gather needed data, conduct analy-
ses, and hammer out its recommendations.

On the more technical side, one of the most difficult tasks the Committee faced was reaching agreement on who would be included in the category of "non-specialist." A sizable majority of undergraduates majoring in the sciences eventually do not pursue a career in science, thus entering the ranks of non-specialists in a special sense. Scientists must also judge the wisdom of the decisions of political leaders in scientific and technological matters in areas often outside their professional specialties: hence, they are non-specialists at times themselves. The Committee concluded, however, that it could properly focus its attention on those individuals enrolled in undergraduate degree-granting programs in two- and four-year colleges and universities who do NOT major in the biological or physical sciences, mathematics, engineering, or the health sciences. The special focus of attention has been on persons preparing to enter business, journalism and communications, education, theology, law, and other non-science professional fields.

"Science education" was limited to the undergraduate study of the biological and physical sciences, technology, and mathematics--including the computer sciences. These fields most often constitute the "natural sciences" component of the distributive or breadth requirements of the general education model adopted by 90 percent of our colleges and universities today. Despite their importance to the advancement of science knowledge, the social sciences have not been included in this assessment of science education for non-specialists. Their treatment in non-specialist undergraduate education should be the topic of a separate report.

The Committee's data-collection plan proved to be workable, but ambitious, given the constraints of time and resources available:

● August 1980-April 1981: Interviews with a sample of science educators who had received support for course-content improvement projects from either the federal government or the private sector.

● September 1980-March 1981: Review of the science education activities of several key federal agencies. Preparation of a paper by Philip Ritterbush reviewing the history of science education of non-specialists.

• November 1980: Two-day invitational hearing in Bloomington, Indiana, allowing students, faculty, and alumni from Indiana University to describe their impressions of the current status of undergraduate science education for non-specialists.

• November 1980-March 1981: Interviews with a sample of non-science professionals regarding their undergraduate experiences in science and their current knowledge needs.

• November 1980-June 1981: Survey of 215 four-year colleges and universities to analyze institutional requirements for the study of the sciences by undergraduates and courses available to non-specialists. A limited set of interviews with science faculty involved with the teaching of non-specialists was also carried out.

• December 1980: One-day invitational conference on past and present efforts to improve undergraduate education of non-specialists in science and technology.

• March 1981: One-day invitational conference with 17 representatives from various non-science professions-- law, journalism, business, theology, public service--to understand what they need to know about science and technology.

In regard to the sources of expertise called on by the Committee, it should first be noted that the individuals who joined in this effort represented a wide variety of fields and sectors. The Committee itself was a diverse group of individuals who brought many different experiences and points of view to their assignment. All have contributed over the years to the improvement of science education--either as innovators in science education for non-specialists or as non-specialists concerned about the quality of science education in general and in their professions in particular. In addition, there was the "extended committee" of participants in the hearings and conference, consultants, and many others who provided information. Altogether we estimate that some 100 persons contributed to the preparation of this report in one way or another.

The Committee worked diligently to come to grips with these complex and far-reaching problems, while recognizing that its efforts would be limited in scope and time. It has striven to prepare a balanced statement of the current situation of college science education for non-specialists and how such education can be improved. The recommendations deal with what the federal government, together with state governments and the private sector, can do to stimulate and assist colleges and universities to give pre-professional students who do not become scientists the science education that they need.

The social problem we have addressed seems very real to us; the need, great; and the solutions, feasible. We submit this report to the attention of agencies of the federal government, the nation's colleges and universities, and thoughtful persons in all walks of life who are concerned with the vitality of the professions whose practitioners--as the report attempts to document--find science and technology of rapidly increasing importance to them.

RICHARD GRAY
Chairman

SUMMARY

This committee has examined the state of undergraduate
science education for those who are non-specialists in
science and concludes that we are presently confronted
by an educational problem of national proportions. Non-
specialists include such opinion leaders as journalists,
lawyers, managers, legislators, theologians, and elemen-
tary-school teachers. The problem is of such magnitude
that it will take the concerted effort of both the public and private sectors to resolve the situation satis-
factorily. Notably the federal government has a role to
play, one that we have tried to define.
     Our conclusion grows out of the fact that the federal
government itself has helped to create both the problem
and the opportunity by assisting to bring into existence
the world of science and technology in which we live.
This nation agrees almost unanimously that, when it is
in the national interest, the federal government should
take steps to solve problems that plague the republic as
a whole. We believe our findings demonstrate such a
need in the present instance. But the federal government cannot do it all. Its efforts must catalyze action
by many performers--first and foremost, the colleges and
universities themselves.
     Our call for action grows out of the Committee's
findings that the nation's colleges and universities,
with a few exceptions, are not doing enough to provide
our future civic and professional leaders with the understanding of science and technology that they need to
function effectively. Our study shows that:

• The historical evolution of college science education has benefited the science major immensely but has
left the non-specialist largely unattended.

• Colleges and universities in general have lowered their science requirements over recent years to the alarming point where the average non-specialist student devotes only about 7 percent (135 contact hours) of a college course load to work in the sciences.

• Within such subminimal requirements, these students are often allowed to choose willy-nilly from an ever-growing cafeteria offering "topics courses" that rarely fit into a well-conceived, comprehensive pattern of education.

• In many cases, those topics courses, which were designed as a response to the student concern for relevancy in the 1960s, have outgrown their relevancy.

• In all too many other cases, those topics courses, as they reach for relevancy, fail to provide students with an understanding of the basic principles of science.

• When students do opt for more traditional introductory science courses, learning often suffers because so many students come to college ill-prepared in secondary-school science and mathematics.

• These students often are subjected to inadequate teaching that stresses dull lecturing more often than exciting laboratory experiments and demonstrations.

Education for the non-specialist is waning at the very moment when history and mankind's ingenuity make the need for knowledge of scientific principles and technology ever greater for those wishing to exert leadership. Unless they take hold of scientific principles, lack of scientific knowledge may very well hold them back from achieving their full potential. Unless professionals master the new technology, it may very well master them.

The Committee's concern rests upon three major contentions. First, enlightened non-specialists are essential to help implement the pluralistic function of democratic decision making about pressing matters of science and technology. Second, knowledgeable non-specialists must serve as opinion leaders in the American political structure to help the public at large understand the complexities as well as the risks and benefits of sci-

ence and technology. Third, well-prepared non-science
specialists can lead the way in their professions more
effectively if they have a command of science and technology.

Our call for a new and deeper understanding of science and technology places special obligations on higher
education. Colleges and universities should ensure that
undergraduate education for non-specialists is an "enabling" process embracing the following goals:

● College science education should enable non-specialists to overcome fears that might prevent them from
launching a lifetime learning experience about science
and technology.

● College science education should enable non-specialists to develop their capacity to engage in critical
thinking.

● College science education should enable non-specialists to know how to seek reliable sources of scientific
and technical information and how to use them throughout
life.

● College science education should enable non-specialists to gain the scientific and technical knowledge
needed in their professions.

● College science education should enable non-specialists to gain the scientific and technical knowledge
needed to fulfill civic responsibilities in an increasingly technological society.

In order to eliminate barriers that prevent colleges
and universities from successfully preparing students to
reach these goals, a number of institutions and agencies
must work cooperatively. The major responsibility rests
with the college and university science faculties. But
others--state governments, private foundations, industry, and the federal government--need to assist. We see
the federal role in all of this as being primarily catalytic. The federal government should help stimulate
action, coordinate efforts across the fifty states and
the many different agencies involved, and serve as a
clearinghouse for exchanging ideas about how to improve
science education for the non-specialist.

We recommend:

1. That colleges and universities find new and additional ways to identify and reward high-quality teaching of science courses for non-specialists. Prizes, sabbaticals, and increased consideration of teaching contributions when tenure and salary decisions are being made should all be a part of a planned incentive program by higher education working in concert with governmental bodies and the private sector.

That the federal government provide a competitive program of grants of $20,000 to $25,000 each to establish model programs in a variety of college settings to explore innovative approaches to evaluating and rewarding college science instructors.

That the president establish and give national recognition to an annual White House Award of at least $5,000 to a teacher who has been selected on a national basis for doing a superior job of teaching science to non-specialists.

2. That colleges and universities that have, in the last two decades, lowered their science demands for graduation, reverse course and raise their requirements. We believe that no less than a total of two one-year courses selected from the biological and physical sciences and mathematics should be required of non-specialists for the baccalaureate degree.

3. That college science faculty restructure introductory subject matter courses and redesign special topics courses to meet the changing educational needs of undergraduate non-specialists. The federal government, together with the private sector, should make financial awards available to realize this goal.

That the federal government fund projects to explore ways to develop courses that emphasize firsthand experience with phenomena, laboratory exercises, and demonstrations that are relevant to the needs and experiences of non-science majors.

4. That colleges and universities provide a forum for scientists and non-science professionals to explore together new directions in science education for non-specialists. Through regularly scheduled faculty meetings, seminars, or retreats, faculty should be encouraged to develop science experiences appropriate to the

educational needs of undergraduate non-specialists. These efforts should be guided by regular consultation with leaders in the professions.

That support for the "Chautauqua series" (p. 69)—a means of stimulating teaching ideas and disseminating innovations for the science classroom—be restored to about $1,000,000 annually by the National Science Foundation.

That the federal government seek out an organization through the National Science Foundation to establish a national directory of teaching innovations in college science education for non-specialists that will be regularly updated and fund it at an appropriate level to create a quality communication link among the nation's teachers of science for non-majors.

5. That colleges and universities extend the use of non-traditional instructional media in teaching science to non-majors in new and possibly more exciting ways. Special attention should be given to the educational potential of the mini- and micro-computers and to such public broadcasting ventures as the Annenberg project (p. 65).

That federal support be made available to evaluate the quality of existing computer-based undergraduate science courses with respect to their potential value to non-specialists.

6. That, in light of the experience of the college science commissions in the 1960s, all professional societies provide more leadership in educational innovation and propagate information widely about new directions in science education for non-specialists. To the extent they require financial assistance, the federal government and the private sector should supplement funding.

7. Finally, that the federal government focus its efforts to oversee improvement of undergraduate education for non-specialists in science and technology by establishing a vigorous program in the National Science Foundation for this purpose. The Foundation should be given responsibility for establishing a clearinghouse for monitoring the diverse activities of the various federal agencies operating in this area. Most important of all, we urge the Foundation to assume this leadership role with considerably more dedication and aggressive-

ness than it has exhibited heretofore toward advancing science education for non-majors.

That the present percentage of the NSF science education budget devoted to the improvement of education of the non-specialist--estimated at 2.5 percent--be raised to something in the range of 5 to 10 percent as more reasonable.

Science for Non-Specialists:
The College Years

1

A RATIONALE FOR THE IMPROVEMENT
OF THE COLLEGE SCIENCE EXPERIENCE

According to legend, Destiny came down to a remote South
Sea island one day in a cloud of doom and warned the
inhabitants that a great tidal wave was coming. Then,
as a test of the natives' ingenuity, Destiny asked three
representative leaders what each would do about the in-
evitable inundation. The first respondent was a hedo-
nist. He answered that he would gather together his
fun-loving friends and have one last party to enjoy as
much wine, women, and song as possible before dawn, when
the great wave would end their pleasure forever. The
second was a mystic. He said that he would seek out the
most pious people he could find and make a pilgrimage to
the sacred groves to pray for deliverance. The third
respondent was a sage. She stoically explained that she
would search the island over for the wisest men and
women she could find, and together they would sit down
and discover how to live underwater.
   Americans trying to cope with the real world complexities of the 1980s face tidal forces of their own kind
and making that very well could inundate them, too. As
in the case of Destiny's South Sea visitation, sages
once again are required. In this instance, we need wisdom to address many of the problems that result directly
from, and may be alleviated by, science and technology.
For example, we need more data and better-informed leaders to deal with the regional and national debates that
periodically erupt over such perplexing issues as genetic engineering, nuclear fallout, the use of pesti-
cides, and the proper employment of life support machines.
   Many new research endeavors and technological innovations can bring negative impacts to some segment of
society, no matter how positive their overall conse-

1

quences might be. These negative impacts may result
from threats to our moral and ethical beliefs, from in-
equitable distribution of the products of new technology, or from decisions to commit sizable allocations of
tax dollars to innovations or space exploration rather
than to education or social welfare programs.

In short, science and technology are often dichoto-
mous forces. They hold potential for good as well as
for evil. They can lead to progress or to regression.
For example, "splitting the atom" has opened the way to
unlimited new sources of energy but at the same time has
provided militarists a heinous weapon of unprecedented
horror. The invention of the combustion engine has
brought mankind hitherto unknown comfort and speed in
transportation and yet has resulted in machines that
clog our lungs and our eyes with a social disease called
pollution. In the final analysis, whether science and
technology bring us progress or decay depends on how
scientific and technological developments are implemented.

To guarantee that such implementation takes the right
direction, wisdom is needed to ensure that science and
technology meet society's basic requirements with a minimum of attendant negative impact. In decision after
decision, we necessarily will have to weigh the benefits
against the risks, although we sometimes understand the
benefits better than the risks, which aren't as clear.
As a nation, then, we will need wise decision makers,
administrators, opinion molders, and leaders of every
variety to guide us successfully through the debate to
reasonable action.

This means that the nation's colleges and universities must prepare leaders capable of bringing facts and
wisdom to the public forum. Most of these individuals
are college graduates who have little or no background
in science and technology. In light of that deficiency,
this report focuses on the state of college science education for undergraduates who are not science majors or
studying engineering. It first examines the problem,
then considers steps needed to strengthen science education for non-specialists, and finally recommends an appropriate role for the federal government to play in
this mission.

3

> Scientifically enlightened citizens are
> essential to help implement the pluralistic
> function of democratic decision making about
> science and technology.

Science and technology, like tidal waves, can have dev-
astating effects when they build up overwhelming force
in shallow water and narrow confines. With scientific
and technological problems, the shallows can result from
decisions fashioned by people lacking depth of knowledge. Making decisions on too narrow a basis of expertise can have equally deleterious effects. That is why
pluralism--or the involvement in policy formation of a
diversity of factions on the social-intellectual spec-
trum--is so basic to the American way of life. As theo-
logian Reinhold Niebuhr observes: "Man's capacity for
justice makes democracy possible; but man's inclination
to injustice makes democracy necessary" (Niebuhr,
1953). Niebuhr's dictum applies as much to the scientific realm as it does to politics. Just as checks and
balances among the three branches of government help
avoid injustice in conducting civil affairs, so counter-
arguments from various constituencies in the social-
intellectual milieu help avert injustice in forming science policy.

In carrying the argument even further, Washington
attorney Harold Green, who has written thoughtfully on
matters of genetic control and public policy, contends
that past discussions of those issues suggest that broad
participation is essential in treating such issues.
Public controversy needs to be stirred up, he believes,
because scientists generally represent only a narrow
spectrum of social values. Green argues that science
policy, like tax policy, should be subjected to the democratic process--including bruising political debate.
He continues: "I do not see anything that is inherent
in science that ought to distinguish it from any other
aspect of our society in terms of the operation of the
political process. Everything else is subject to the
adversary process and debate; why not biomedicine?"
(Green, 1976).

This is not just lawyer talk. June Goodfield, senior
research associate at Rockefeller University and adjunct

professor at Cornell University Medical College, insists
that scientists must not "commit the cardinal error of
ignoring the fact that their profession sits squarely
within the social matrix." Goodfield declares: "It is
this social matrix that helps determine the directions
in which science will go and the social acceptance of
its fruits, yet apparently scientists have still not
fully realized that the contemporary social matrix demands that the desires of ordinary people must be re-
spected and acknowledged" (Goodfield, 1981).

John Ziman, Henry Overton Wills Professor of Physics
at Bristol University, is equally insistent in calling
for diversity in setting scientific and technological
policy:

It is the wisdom of the pluralistic society to
doubt the competence of any authority to choose
wisely on behalf of every citizen. It is not so
much that they cannot be trusted not to feather
their own nests; it is simply that the questions
debated . . . are seldom correctly posed. They
refer far too much to what is technically possible
or technically optimal, rather than what is so-
cially desirable. This is perfectly exemplified
by the history of the Concorde project, where the
advisory committee seems to have been dominated by
engineers and accountants, rather than by potential passengers or by non-passengers residing near
large airports. . . . (Ziman, 1976)

The proper antidote to what Professor Ziman calls
"the poison of technocracy" is participatory democracy.
After all, the old sage in the introductory fable sought
wisdom wherever she could find it, not just from underwater experts. There is a persistent and troubling difficulty to participatory democracy, however. It is easier to chart than it is to implement. As Walter
Lippmann has adroitly observed about the pluralistic
society: "No amount of charters, direct primaries, or
short ballots will make a democracy out of an illiterate
people. Those portions of America where there are
voting booths but no schools cannot possibly be described as democracies." Lippmann called for "a nation
vastly better educated, a nation freed from its slovenly
ways of thinking, stimulated by wider interests, and
jacked up constantly by the sharpest kind of criticism"
(Lippmann, 1913, in Rossiter and Lare, 1963). Not all

citizens are capable of participating in this tough process. But at a minimum there must be a sufficient number and diversity of opinion leaders taking part to guarantee that all important points of the social-intellectual spectrum enter into the analysis and refinement of forming scientific and technological policy. Otherwise, our scientific establishment may either stagnate or run rampant beyond humanistic control.

These concerns place a very heavy burden upon our colleges and universities to prepare bright young men and women majoring not only in the sciences but in other disciplines as well. We must have a variety of intelligent non-scientists who are capable of carrying on knowledgeable and critical discussion about scientific and technological issues if we are to arrive at policies about space, energy, and the like through a pluralistic process. To keep democracy operable, then, our system of higher education is going to have to turn out lawyers, journalists, ministers, politicians, and other professional leaders who are capable of engaging scientists at least partially on their own terms. Otherwise, we may confirm the fears of those who claim that widespread involvement in decisions concerning scientific issues will paralyze research and technological advancement. Otherwise, we may find ourselves members of a society in which only a small band of experts influence an elite group of politicians in order to work their self-centered will.

Thus, to help ensure the survival of pluralism in American democracy, we believe it is imperative that the federal government play a strong role in encouraging quality science education for college non-specialists. To play this role, then, is squarely in the national interest.

Knowledgeable college graduates must serve as opinion leaders in the political structure to help the public at large understand science and technology.

This participatory process we have been discussing functioned exceedingly well in the early years of the

United States government near the turn of the eighteenth
century. Sovereignty was limited to probably no more
than 10 percent of the population and was wielded by men
who for the most part were liberally educated about science as well as most other important subjects. In fact,
a number of early American political figures not only
could debate science but practiced it. Benjamin Franklin, for example, was elected to the Royal Society of
London and was awarded its Sir Godfrey Copley Gold Medal
"on account of his curious experiments and observation
on electricity" (Van Doren, 1952).

Science was not yet a professional preserve, nor was
it very specialized. As a result, men who could afford
the leisure and the education were able to become pro-
ficient in scientific matters. Historian Verner W.
Crane explains:

> In America, as in Europe, clergymen like Cotton
> Mather, officials and gentry like Cadwallader
> Colden, and numerous physicians could set up as
> spare-time natural philosophers; the claims of not
> a few were recognized by election to the Royal
> Society. In Philadelphia, even tradesmen and ar-
> tisans advanced in a few years to new frontiers of
> experimental science. (Crane, 1954)

As the United States moved from what originally was
an agrarian society into the industrial age, then into
the atomic era and ultimately into the space age, fundamental changes occurred that gradually upset this happy
relationship between science and politics. Science, and
before long technology, grew increasingly sophisticated
and specialized. At the same time, the body politic
broadened as suffrage was extended beyond the old landed
aristocracy to the rising merchants, small farmers, the
working classes, women, and finally minorities. The
United States also expanded from a country of 13 states
along the Atlantic seaboard into a 50-state nation of
more than continental proportions. The result is a
widely dispersed populace who have equally wide dispar-
ities in their abilities to cope with the pressing issues of science and technology--issues that are as complex as they are necessary for existence in modern life.

Largeness, in and of itself, is enough to strain the
democratic relationship between science and politics.
As historian Carl L. Becker notes: "It is a striking

fact that until recently democracy never flourished except in very small states—for the most part in cities."
True, both the Romans and the Persians accorded a measure of self-government to local communities in their
empires but, Becker insists, only on purely local matters. He explains: "In no large state as a whole was
democratic government found to be practicable. One essential reason is that until recently the means of communication were too slow and uncertain to create the
necessary solidarity of interest and similarity of
information over large areas" (Becker, 1941). This
partially explains why the Greek city-states, where democracy first took form, were kept small. As much as
geography, it may have been some political instinct
among Greeks that told them a state necessarily should
be a natural association of people bound together by
traditions and obligations based on common knowledge and
understanding. When the Greeks began to build an empire
that transcended the city-state, they soon lost many of
their democratic tendencies.

One way to overcome the difficulty of applying democracy to a large industrialized nation such as the United
States is to create a strong sense of community in the
Greek tradition through modern means of communication.
In this fashion, political kinship can be built on a
basis of understanding science and technology as well as
other issues vital to contemporary life. As Carl Becker
explains, the reason the republican form of government
has survived into the twentieth century in the United
States is that "the means of communication, figuratively
speaking, were making large countries small" (Becker,
1941).

These means of communication flow from a variety of
institutions serving modern democracy—the media,
churches, and public schools, to name a few. All disseminate information. All help form public opinion.
All depend on college-educated professionals to operate
effectively. And none will be able to lead public opinion intelligently unless those professionals receive a
meaningful college education in the sciences.

It is important that enlightened opinion leaders rep-
resenting a variety of interests be involved in helping
build a broad base of understanding about scientific
matters. William G. Wells, head of the Office of Public
Sector Programs for the American Association for the
Advancement of Science, explains this need:

Why is it important for us to have science broadly based? Well, all citizens have to be involved in the great debates of our society today that increasingly involve science and technology in some manifestations.

For example, today we are in the middle of a virtually nationwide debate on the whole subject of creationism versus evolution. It is in the courts. It is an issue that is not going to go away. . . . I would argue that this is not a debate solely for the biologists or the anthropologists. This is a debate for all of us because it does affect the fundamental basis, the fundamental nature, the fundamental understanding of what science is about, the role that science plays, the ways of scientific thinking, and the meaning of scientific evidence. (National Research Council, 1981c)

In the decades ahead, the public at large will be required to make difficult choices regarding the extension of scientific research and technological development in the face of limited resources. In its search for the good society, what trade-offs will the public be willing to make? Will voters allow state and local governments to condemn huge tracts of land in order to erect banks of solar collectors to transform the sun's energy for urban use? Will citizens accept the costs of controlling pollution in order to enjoy the benefits of clean air and water? Will the public allow officials to commit billions of dollars to establish space colonies while one-fourth of the earth's population lives in poverty?
Reasonable resolution of these and other problems of equal magnitude will depend in large part upon the caliber of the public opinion leaders who emerge from the nation's colleges and universities in the years ahead. Will these leaders—most of whom will not have majored in one of the natural or physical sciences—be able to help the public distinguish sense from nonsense? Will these leaders know where to go to help the public find answers to perplexing questions in an era that has seen an almost overwhelming expansion of knowledge? Will these leaders be capable of leading the public to reliable experts when scientific celebrities wage campaigns in direct conflict with one another? To a significant

degree, answers to these questions of public trust will
be determined by the quality of science education for
non-specialists provided on the nation's campuses.

Well-prepared non-specialists can assume
leadership in their professions if they have
a command of essential scientific concepts.

Non-specialists have obligations to themselves and to
their professions in addition to their public responsibilities. As we move further and deeper into an age
where computers and other technological inventions touch
our everyday lives, meeting these obligations requires
better and different college preparation in the sciences
than has been offered professionals in recent decades.

The field of journalism provides a striking example
of this need. Within the last 20 years the profession,
after having been relatively static in terms of technology for nearly a century, has undergone a technical
revolution. Invention after invention has changed
nearly every process of journalism, from news gathering
to news delivery. This shiny new information technology
includes computerized word processing, offset printing,
data retrieval systems, electronic interoffice word and
picture transmission, laser printing, the electronic
camera, and computerized home delivery of newspapers.

These developments are so sweeping and so revolution-
ary that they are bringing about fundamental changes in
the very power structure of journalism. In effect, the
traditional power structure that has been in place for
the last 100 years came about with the invention of the
Linotype and stereotyping in the late 1880s. At that
time, editors lost control of newspapers because they
failed to understand and master the technology. The
production people who managed the back shop began to
gain power because they understood the machinery and
what it could do. Publishers, therefore, began to give
more weight to the technical production people rather
than to editors in arguments about such matters as
whether there would be front-page remakes, extra edi-
tions, or additional pages.

Now editorial people are in a position to move back to the top of the journalism power structure. Managing editors, if they understand technology, can regain control of the editorial product by telling the computer people what to do and how to do it. They need not let computer salespersons just impose on them.

In the early stages of the application of computers in journalism, companies simply picked up computer systems that had been designed for banks and airlines and applied them to newspapers. Because editors had no knowledge of how computers operate, they took a lot of nonsense from programmers who would say "You can't do that" when indeed they could do that. Computer experts could design keyboards and control systems to do what journalists wanted done. If journalists had known enough, they could have insisted that it be done. So if journalists are going to control their own destiny and exercise proper professional authority over editorial decisions, they will need a fundamental knowledge of science and technology. This example, drawn from one professional field, could be replicated in almost every other field discussed in this report.

The indicators are clear, then: History and mankind's ingenuity are taking us deeper and deeper into an era in which familiarity with scientific principles and technological know-how is becoming a commonplace requirement for those wishing to exert leadership. Unless professionals master the new technology, it may very well master them. Unless they take hold of scientific principles, lack of scientific knowledge may very well hold them back. More and more professionals are going to need such scientific knowledge in addition to quantitative analytical skills and knowledge of the computer just to perform their day-to-day work. How deep this knowledge should be and how it should be provided are the subjects of this report.

More important, however, society is going to look increasingly to professionals for leadership in working with scientists to make sure that science and technology are harnessed on behalf of progress and good, rather than regression and evil. Unfortunately, America has been much more successful in creating a remarkably complex and powerful technological order than it has been in avoiding the social and psychological problems that accompany technology when it goes to excess (Hannay and McGinn, 1980). Professionals such as clergy, lawyers, journalists, and businesspersons could be instrumental in helping the United States adjust its technical-economic growth to keep pace with its concern for human values.

To do so will require unprecedented creative wisdom--of the kind called forth by the sagacious woman trying to offset a tidal wave. Securing wisdom in the 1980s will be much harder than in the mythical situation, however. It will take disciplined minds that have given more than passing attention in college to science and technology. Educating those minds will require a dedicated commitment on the part of higher education, with help from the federal government, to provide non-specialists a better college science experience than they have heretofore been receiving. The challenge is to help bright young non-specialists overcome initial anxieties and other obstacles so that they gain a facility to engage in critical thinking about science, a facility that will, with periodic efforts of renewal, serve them and society through the rest of their professional careers.

2

THE UNIVERSITY'S OBLIGATION TO EDUCATE FOR LEADERSHIP IN A SCIENTIFIC AND TECHNOLOGICAL AGE

The function of responsible opinion leaders is to help clarify issues, explore alternatives, and guide the various constituencies of democracy to reasonable conclusions. The public at large in modern democracy does not have to deal with the specifics of science, as citizens of the early republic did or as some professional leaders still must do in the interest of maintaining pluralistic decision making. The general public does, however, need to give mandates and elect officials who will carry out scientific and technological policies for the public good.

To a significant degree, improved protection of the public interest will be determined by the quality of science education for non-specialists provided on the nation's campuses. As college graduates move into positions of professional leadership and influence public opinion on scientific and technical matters, it is critical that consideration be given to the extent college science education contributes to their future role as opinion leaders. In light of this important obligation, we believe colleges and universities should ensure that undergraduate science education for non-specialists be an "enabling process" embracing several goals. This chapter will describe what this process should be and will provide examples of how the goals can be met.

> College science education should enable
> non-specialists to overcome fears that might
> prevent them from launching a lifetime
> learning experience about science and
> technology.

All indications are that many of those students entering
college today but not planning to major in science are
interested in science and understand how studying it
would benefit them. They are, however, either intimi-
dated by formidable introductory courses for science
majors or dissatisfied with what they see being offered
in courses for non-scientists (National Science Foundation, 1980; Mallow, 1981).

One student majoring in elementary education told the
Committee at its hearings at Indiana University:

> When I first came to IU and found out I had to
> take 18 hours of science, I nearly had a heart
> attack! In high school, I did the bare minimum to
> graduate. . . . I think I had two credits of biol-
> ogy and then I had two credits of math. . . . So I
> had no background at all to come to college and
> take all these classes. (National Research Coun-
> cil, 1981a)

This is not the isolated opinion of one undergraduate. Similar complaints are directed hundreds of times
every year to student advisors across the country, according to the Committee's survey of science teachers
and the experience of several Committee members who are
long-time science teachers themselves.

In testimony before the Committee, Christine Harris,
director of the Consortium for the Advancement of Minorities in Journalism Education, had this to say about
students' reactions to science requirements in that
field:

> Many journalism students shy away from studying
> science and math. They are afraid of it and in-
> timidated by it. Every one of the educators I
> talked with noted that a large number of students
> take as little math and science as they have to to

get out of college. (National Research Council,
1981c)

Unless students overcome their fear of science, they
will not be likely to learn. One approach to counter-
acting the fear of science expressed by entering college
freshmen is being developed by Hans Andersen at Indiana
University for students majoring in elementary education. With partial support from the National Science
Foundation, the "Indiana Model" for training teachers
attempts to integrate the "learning" of science with the
"teaching" of the subject (National Research Council,
1981c). A three-part sequence offered simultaneously
within a semester couples a content course in a science
area such as physics, biology, or earth sciences with a
teaching methods course. Students are then sent out to
elementary school classrooms in the Bloomington area to
practice teaching the science they have just studied.

This approach has had a fascinating, positive effect
in dispelling initial fears of science expressed by students entering college. Actually having to translate
the science for someone else, namely elementary school
children, seems to bring it down to a level where fear
evaporates, where all that "theoretical kind of bookish
stuff goes" (National Research Council, 1981a).

As Committee member H. Richard Crane commented at the
Indiana student hearings:

> The kind of enthusiasm produced by this science
> sequence suggests that other parts of the univer-
> sity might be able to learn something from the
> education department. A sequence of science
> courses like these might be applicable to more
> people than just those in elementary education.
> Students in English or in history or something
> like that might like to learn science this way.
> (National Research Council, 1981a)

The apparent success of Andersen's program suggests that
students who will have responsibility one day for communicating information about science and technology--
including those majoring in journalism, education, and
theology--should be encouraged to couple the learning of
science with practice in communicating about science.
For example, reporters on student newspapers might be
encouraged to write feature articles about science
developments on the campus. Students studying theology

might be asked to preach sermons on science and society in homiletics courses.

College science education should also motivate the student to want to learn more about science, to follow new ideas, and to understand what scientists are talking about, even after graduation. The non-specialist, after all, will spend on the average at least 12 times the length of undergraduate education pursuing a living. One lawyer interviewed by the Committee put the value of science courses in later life this way: "[College has] generally influenced my reading habits in science" (National Research Council, 1981c).

Through creative application of teaching talents, college professors can motivate non-specialists to take a genuine interest in science. In doing so, they should enable students to discover that they are able to (1) understand scientific phenomena, (2) extend their own knowledge of scientific things, (3) derive satisfaction from learning about scientific things, and (4) continue to develop their knowledge and critical interpretation of new information during the rest of their lives. Many professors provide this type of experience now; many more need to assess why they are not more successful in doing so.

It is important that non-specialists develop some interest in the way knowledge is acquired if they are to increase their stock of scientific information in the years following graduation from college. Broad under-standing and a grasp of the basic principles are needed. There should be no hurry to rush undergraduates into a detailed analysis of the natural sciences. Instead, undergraduate non-specialists should be able to complete their college science experience with some modest sense of accomplishment. Even more important, they should come to realize that the further study of science is within the range of their abilities. Most important of all, they should have a college experience that allows them to discover that science can be fun and exciting. Once fear has been overcome, the student engages in a college science experience that can truly sharpen his or her critical thinking abilities.

> College science education should enable
> non-specialists to develop their capacity to
> engage in critical thinking.

By "critical thinking," we mean the ability to grasp information, examine it, evaluate it for soundness, and apply it appropriately. Therefore, properly designed undergraduate courses in the sciences should impart to the learner some of the cognitive strategies employed by the scientific investigator when engaged in the act of inquiry. The thinking skills conveyed by such courses have the potential to be of both general and specific value to the non-specialist. In general, science education can assist the student to apply a well-trained mind to a wide variety of endeavors. Whether engaged in business, journalism, law, or teaching, the individual who can sort sense from nonsense is one of the "most critical of our national assets, among the scarcest and the most valuable of our national resources" (President's Science Advisory Committee, 1959).

Some would argue, quite justifiably, that science is not unique in providing the student with this sort of acumen. As a field of inquiry, however, science does have a specific contribution to make to critical thinking. Exposure to science--its corpus of knowledge, its vocabulary, the nature of its investigative methods, its limits, and its potential--has special benefits. It can prepare the non-specialist to question scientific pronouncements, to suspect shoddy research, and to identify fraudulent scientific claims.

Opportunities for using a little scientific judgment in everyday life are everywhere: Can the position of stars at birth truly determine one's fate? Is it worth spending family savings to travel to Mexico in search of a magic cure of cancer through Laetrile? Should we have an egg a day as our mothers admonished us or believe the latest research claims about cholesterol? If a particular toothpaste reduces cavities, is it the vigorous brushing that is responsible or the fluoridation treatment or both? Because we no longer see the pollution emanating from smokestacks, does that mean that a factory has succeeded in making the air clean? Because we see a cloud rising from a smokestack, does that mean the air is polluted? These are some of the issues that require critical questioning and judgment. Sound courses in science can help build that judgmental power.

Numerous individuals testifying before the Committee described how the concepts and methods of science can be applied to non-science fields. In law, for example, Lee Loevinger, a practicing attorney with the firm of Hogan and Hartson in Washington, D. C., and vice chairman of the Science and Technology Section of the American Bar Association, drew clear parallels between the role of observation in scientific inquiry and the way judges and lawyers attempt to secure a reliable data base through the litigation process.

Experience with science, properly presented, can soon awaken individuals to the beauty and utility inherent in the scientific data-gathering process. One graduate student, who specializes in science writing, described for the Committee her excitement in discovering the meaning of causality. As a result of laboratory work, she learned to think in terms of "probability," "reliability," "validity," "experimental groups," and "control groups." She explained:

> I was fascinated because my thinking had been uncontrolled before and I hadn't realized it. . . .
> I was accepting as cause and effect things that weren't causally related at all. (National Research Council, 1981a)

Other students described their surprise and delight in finding that the scientific way of thinking has broader application. For example, a speech communication major told the Committee how science and mathematics courses sharpened her analytical abilities in argumentation:

> I have had the introductory biology class for majors. I have had 10 hours of calculus at the 200 level. I have had the introductory computer science class for majors. And I have had a couple of classes in abstract logic. . . . I think that the biggest thing that these classes did for me was to give me analytical tools, not necessarily factual data. I know that from my math background throughout high school and college just the concept of taking a theorem and trying to prove it helped me a lot in speech and argumentation. I

found it much easier just to pick out what is
wrong in arguments. (National Research Council,
1981a)

The same student explained to Committee members how
business, political science, and her own field of speech
communication have borrowed from biology the concept
that organizations act like organisms and can thus be
studied as whole systems. She observed:

I have found that other areas of speech communication often use natural sciences as a paradigm to
help explain what is going on. I know a school of
thought right now in the social sciences is the
so-called "systems" theory school. It is big in
business school. It is big in speech communication, especially when you are studying group
interaction and organization communication. I
know they use it to study the organization of Congress.

The whole paradigm was taken from biology, saying
organizations are like organisms. Although I had
no factual data from biology that specifically
helped me understand this theory, I know that
having biology classes gave me a better appreciation for what the systems school is. (National
Research Council, 1981a)

The study of science can also help students learn to
analyze claims about so-called "truth" and distinguish
more intelligently between arguments that are presented
in black-and-white terms. The graduate student in journalism who appeared before Committee members gave re-
vealing testimony on this point. She recalled how she
had read recently in a newsletter from the Institute for
Public Information how an ABC news executive had explained away the public's increased interest in science
news. He had claimed that the public needed a sense of
certainty because the world has grown so "iffy and
changing" and that science provides definitive answers.
She exclaimed:

I thought to myself, "Oh, no!" This is a popular
misconception on the part of a lot of people, including working journalists, that science provides
definitive answers. That just reinforced my

feelings that we ought to be teaching people more
about the process of science. That science is
dynamic; what is true today is not true tomorrow.
(National Research Council, 1981a)

Numerous modern authors have found the concepts of
science appealing to their dramatic instincts. W. Somerset Maugham, for example, has recalled how his medical education provided him with a rudimentary knowledge
of science and the scientific method that he "embraced
with alacrity":

It was a very limited knowledge, for the demands
of the curriculum at that time were very small,
but at all events, it showed me the road that led
to a region of which I was completely ignorant. I
grew familiar with certain principles. The scientific world of which I thus obtained a cursory
glimpse was rigidly materialistic. . . . I was
glad to learn that the mind of man (himself a
product of natural causes) was a function of the
brain, subject like the rest of his body to the
laws of cause and effect, and that these laws were
the same as those that governed the movements of
star and atom. (Maugham, 1938)

It should be clear, then, that science can not only in-
form the intellect and sharpen reasoning abilities, but
fire the imagination as well. Most important of all,
however, the study of science--when properly done--can
lead students to a life-long attitude that prompts them
to examine data, issues, and opinions with a critical
eye. Students who have learned to explore fundamental
causes, think out theorems, and question scientific
findings rather than merely memorize facts by rote are
prepared to evaluate critically the world of science and
technology. They are also prepared to deal with the
growing complexities in general that beset us in the
twentieth century.

> College science education should enable
> non-specialists to know how to seek reliable
> sources of scientific and technical
> information and how to use them throughout
> life.

Learning to think critically in college is not enough. The college experience at its best should prepare students to acquire analytical skills and an ability to locate the information with which to think critically throughout life. For the typical undergraduate non-specialist, science lectures will have occupied only about 7 percent of total undergraduate course work by the time the baccalaureate degree is conferred (see Chapter 3). This means that most graduates at present leave college with an understanding of science based on an average of 135 contact hours of formal instruction, out of a 4-year total of about 1,860 contact hours.

If those 135 contact hours were spent simply exploring the latest findings of science, the non-specialist's knowledge of science would be rendered obsolete in just a few years. The rapid change and growth of scientific knowledge in the past three decades alone suggest that college science education for the non-major must incorporate a different educational strategy than education for the future scientist. The scientist has years to discover the ongoing nature of scientific inquiry. The non-scientist has only a few courses. Science is another form of continuing human inquiry, and the base of knowledge will change with time. The non-specialist should be prepared for further encounters with scientific information and should know which specialist to call upon to deal with a given matter.

Journalists, lawyers, businesspersons, politicians, and general citizens alike are at the mercy of individuals who correctly or incorrectly are called upon--or seek--to speak as experts in the area of science and technology. How is the non-specialist to know if the individual is making sense? Because a scientist encourages caution in accepting the conclusions he or she has reached as a result of research, should the public dismiss the claims of that scientist in favor of the claims of a scientist who is uncompromisingly certain about his or her facts?

The episode at the Three Mile Island nuclear power plant provides an interesting example of the challenges that confront the journalist who attempts to arrive at the "truth" behind a story. In a starkly dispassionate analysis following the incident, the President's Commission on the Accident at Three Mile Island instructed the Task Force on the Public's Right to Information to assess the way in which public information officers and journalists served the public's information needs. The task force pointed out that the information about the accident had a "significant bearing on the capacity of people to respond to the accident, on their emotional health, and on their willingness to accept guidance from responsible public officials" (President's Commission on the Accident at Three Mile Island, 1979). The task force concluded that neither public information officials nor journalists served the public's right to know:

> Perhaps the most serious failure in the planning stage was that neither the utility nor the NRC (Nuclear Regulatory Commission) made provisions for getting information from people who had it . . . to people who needed it. . . . Given this confusion among sources, and given that reporters are almost entirely dependent on such sources for their information, it is not surprising that news media coverage of the accident in the first few days was also confused. (President's Commission on the Accident at Three Mile Island, 1979)

Such confusion confronts the field of law, too. Increasingly complex issues are being presented in the legal system for which there is little to guide lawyers or judges in making decisions about the validity of what is portrayed as scientific evidence. An attorney preparing for a trial must locate and examine witnesses well before the trial and learn as much as possible about the specific scientific issues in question (Thomas, W., 1978). There is little control, however, over who is willing to come forward as an expert witness or his or her reliability to serve as a source of information. It is not clear that the legal methods employed in court are adequate checks of reliability. Lee Loevinger put it to the Committee this way:

> Cross-examination is still regarded as the best test of truth in modern trials. However, there

are virtually no empirical data validating the
technique and it is today accepted mainly as the
equivalent of a medieval ordeal. (National
Research Council, 1981c)

In summary, the college experience should enable non-
specialists to launch a lifelong quest for knowledge and
understanding. It should give them an understanding of
science that will prepare them to (1) seek reliable
sources of current information, (2) look beyond the content of scientific claims to the care with which the
scientist has framed the statement, and (3) ask the
right questions. This is perhaps the most challenging
goal in teaching science to undergraduate non-specialists. It requires careful planning of the limited time
a science teacher has available to interact with students. It requires the opportunity for discussion and
the creative use of the many examples of reliable--and
unreliable--sources today. It is an investment during
the college years for a life that increasingly will be
affected by science and technology.

College science education should enable
non-specialists to gain the scientific and
technical knowledge needed in their
professions.

Non-specialists should have within their reach the special scientific and technical knowledge needed to carry
out professional activities in an age of rapid scientific change and technological development.
   To understand what non-specialists need to know about
science and technology after professional training is
complete, we convened a one-day invitational conference
on science and the professions. This was supplemented
by individual interviews with approximately 30 non-science professionals in a variety of occupations and work
settings--lawyers, journalists, business managers, theologians, and others. The information we collected suggests that non-specialists more and more are having to
come to grips with scientific or technological concepts
and knowledge as they carry out their professional activities.

A lawyer, for example, working in the Environmental
Enforcement Section of the Department of Justice assists
in coordinating the enforcement of the Clean Air Act and
the Clean Water Act, among other environmental laws. In
these cases, the Department of Justice brings law suits
on behalf of the Environmental Protection Agency (EPA)
against polluters in violation of federal statutes. She
explained:

> I must determine whether the legal standards have
> been complied with. I must determine whether the
> difference between the standards established by
> EPA and the level of emissions is statistically
> significant. I must also determine what methods
> of pollution control are available to the industry
> in question, what the company is doing to control
> emissions, what it can do, and what it would cost
> them to do more than they are doing at the present
> time. (National Research Council, 1981c)

Another example of how technical federal regulatory
law has become in recent years is provided by the comments of an attorney working for the Common Carrier Bureau of the Federal Communications Commission (FCC).
The FCC regulates such conglomerates as American Telephone and Telegraph and Western Union, authorizes satellite positions in space, and regulates the use of car
telephones and telephone rates. Companies apply to the
FCC to operate as common carriers. They must abide by
regulations and rules established by the FCC. While the
judgment is a legal one, there are often engineering
aspects that must be considered. For example, an attorney who handles such cases told the Committee:

> A company which applied recently wanted to operate
> a facsimile service, transmitting hard copies of
> materials between the 48 contiguous states and
> Alaska. A question was raised with respect to the
> band width of the channels which could be used by
> this commercial operator. As it turns out, the
> band widths of certain channels are not sufficient
> to take the type of information this service would
> transmit. I had to approach the engineers at the
> Bureau to work through the technical details of
> the proposal vis-a-vis the regulations established
> by the FCC with regard to band width use.
> (National Research Council, 1981c)

The modern attorney working in the areas of regulatory law, law relating to computers, patent law, space law, technology assessment, or corporate law increasingly confronts situations directly involving scientific and technological information. A professor of law at Indiana University, who started out in college as an engineering major, explained to the Committee how important a command of science is in estate planning:

> As soon as you get into any sort of sophisticated personal or financial planning, you immediately have to deal with concepts and with instruments that have their basis in the same sort of things that I was dealing with in my first year of engineering courses; that is, how a computer operates, what its capabilities are, and basic number theory. Believe it or not, there are some mathematical concepts hidden in the Internal Revenue Code, and to be able to not only extract those but transmit them to your client is absolutely essential. (National Research Council, 1981a)

He went on to say that one of the chief problems in dealing with estate planning is that the computer is replacing "the traditional avuncular attorney." He declared: "More and more, it is the projection that appears on the CRT that tells the story, rather than some maxims that have been tossed around the office for the last 50 years."

Some lawyers would call for an even broader understanding of scientific principles. Lee Loevinger advised the Committee that lawyers, as well as legislators and other professional intellectual workers, need a grasp of the following principles to perform at what he calls a completely competent and adequate level:

> First and foremost, for example, is the principle of parsimony, Occam's razor, which requires an economy of concepts, not of money. Imagine the revolution in government, if you will, that would occur if this principle were understood and respected by the personnel laboring in Washington.
>
> Next . . . contrary to popular impressions, science is not a body of certain, immutable, fixed and precise principles. It is rather . . . a body of probability statements. Indeed, the whole con-

cept of probability is at once one of the most
fundamental and most elusive concepts in the
fields of both science and law. Both legal and
scientific conclusions are never certainties but
only probability propositions. Yet the concept of
probability, except in its most popular and intuitive sense, is studied and understood by very few
in law. (National Research Council, 1981c)

In the area of business management, corporate leaders
are also finding that they need to keep up with the
rapid changes in science and technology if they are to
succeed in highly competitive markets. The president of
an international television program distribution firm
based in New York City described how he has had to deal
with innovations in technology this way:

Television programs are distributed worldwide by
satellite transmission systems today. In order to
supervise sales, I had to learn about the technical operation of videosystems. What does it
take to get it on the air or recorded? What are
the differences in TV standards throughout the
world? I don't have to know how to operate or to
fix equipment, but I do have to know what goes
into putting a program on the air. I have to know
how satellites operate, and whether the electronic
[TV] standards in one country will permit us to
broadcast a program by satellite to the U.S. (National Research Council, 1981c)

The need for a knowledge of science and technology is
not restricted to individuals working in fields of management or law. Consider the experience of an interna-
tionally recognized landscape architect and regional
planner:

I've planned and designed numerous sites in the
United States and abroad and have conducted many
regional assessments. Projects include: a survey
of the upper northwest quadrant of Colorado for
recreational use; the design of a new town outside
Houston, Texas; the selection of an appropriate
site to locate the capitol of Nigeria; the development of an environmental park in Iran. . . . To
do my work, I must put together teams of experts
from a wide range of fields--the physical, biolog-

ical, and social sciences. I incorporate information drawn from the soil sciences, ecology, an-
thropology, biology, social science systems, etc.,
etc. Everytime I do a project I have to hire a
whole range of specialists. Every situation is
unique. Because it is impossible for me to have
mastered all the information I need to know, I
depend on experts to bring their information to
bear on the problem. (National Research Council,
1981c)

We suggested at the outset of this report that there
is also a significant pool of individuals who are responsible for transmitting information about science and
technology to the general population. As reporters,
teachers, or theologians, these people have important
professional needs to understand science.

Representatives from the field of journalism, for
example, point out that it is important to distinguish
between two categories of reporters when thinking about
their needs for scientific knowledge. These are general
assignment reporters and beat reporters. General assignment reporters will be called on to cover science
stories in areas where they may never have done a story
before. They may have no background in the subject matter and find they "must get it on the spot" (National
Research Council, 1981c). In contrast, beat reporters--
who cover a single topic or field of knowledge--have the
luxury of building up their understanding of an area
like energy, the environment, or the medical sciences.

One individual testifying before the Committee described how Roger Witherspoon, whose "beat" is in the
area of energy for the Atlanta Constitution, goes about
gathering data for an assignment. He obviously has a
need for scientific information, as the following testi-
fies:

Even before he started that beat, one of the
things he did was to visit a nuclear power plant.
In fact, he visited two, one that was completed
and operating and another that was under construction so that he could see how the thing was put
together. He talked to engineers. He talked to
scientists to get an understanding of what nuclear
energy was all about. He did all of that background research before he even started his beat so
that when he started writing those stories, he had

a basic understanding of the subject matter. (National Research Council, 1981c)

Another group of non-science professionals who need good college science preparation is the clergy. LeRoy Walters, professor at the Kennedy Institute for Bioethics at Georgetown University, described for the Committee how clergy have a practical need to understand developments in science and technology in order properly to counsel members of their congregations on the crises of life. He explained:

> For example, the clergy may need to know about genetic counseling centers in the vicinity of their churches or synagogues to refer members of their congregation for expert technical advice. Similarly, members of the clergy may need to have a basic understanding of diseases or probable outcomes of particular illnesses in counseling with the parents of seriously handicapped newborn infants or the adult sons and daughters of seriously or terminally ill parents. (National Research Council, 1981c)

College science education should permit non-specialists like those depicted above to gain the scientific and technical information they need to carry out their professional roles. For some, this may mean a solid grounding in the same introductory science sequences provided for science majors. For example, one individual we interviewed who writes a nationally syndicated column on the environment believes that those students interested in working in a specialty area such as environmental reporting should take courses in science fields related to environmental matters, "including basic biology" (National Research Council, 1981c).

Our concern is largely with the many undergraduates majoring in fields where little effort has been made to date by educators to relate science to the professions. We believe that the scientific community should do much more to offer undergraduate science courses of special value to future non-science professionals. How this can be accomplished depends to a large extent on the resources and talent available at the various undergraduate institutions. We shall offer some suggestions later in this report.

College science education should enable
non-specialists to gain the scientific and
technical knowledge needed to fulfill civic
responsibilities in an increasingly
technological society.

Most important issues in the public arena today involve
science and technology. These include such topics as
nuclear proliferation, the use of chemical additives in
foods, the impact of medical technology on the individual and the family, and energy conservation.

The recent incident in California involving the Medi-
terranean fruit fly provides an interesting example.
Concerned that a federal quarantine would cripple the
state's $14 billion farm industry, Californians had to
decide whether the health risks associated with the aerial spraying of the pesticide Malathion outweighed the
economic consequences of failing to halt the infesta-
tion. The state had already attempted to combat the
threat by confiscating and destroying infected fruit
when it became clear that aerial spraying would be vital
to a successful effort. Governor Jerry Brown at first
resisted the idea of aerial spraying on environmental
grounds but reversed his decision soon after Agriculture Secretary John Block announced it would be necessary to quarantine the California produce. After a period of apocalyptic rhetoric, Californians generally
took the spraying in stride (Wallis, 1981). The wisdom
of the decision to proceed with the spraying has yet to
be determined. Nonetheless, the situation as it has
presented itself provides a dramatic example of the
problems that arise when many legitimate points of view
must be sifted in order to decide upon the appropriate
use of technology--often under pressure of an urgent
need to act.

In another area of public policy, legislation was introduced this year by Senators John H. Chaffee (R-Rhode
Island) and Thomas B. Evans (R-Delaware) that would establish a Coastal Barrier Resources System consisting of
undeveloped coastal barriers on the Atlantic and Gulf
coasts--including barrier islands and beaches, baymouth
barriers, and tombolos. One of the primary goals of the
bill is to prevent new federal expenditures and finan-

cial assistance for construction within the proposed
system.

Federal government meteorologists and traffic engineers have watched uneasily as the population density
along our coasts has nearly doubled over the last 20
years. The rate of urban growth on barrier islands between 1960 and 1976 was four times the national average. It has been estimated that each year between 5,000
and 6,000 acres are urbanized (Chaffee and Evans, 1981).
A number of low-lying areas along the Atlantic and Gulf
coasts have become so densely populated that even with
advance notice, many of their inhabitants could not be
evacuated in time to escape from hurricanes (Flattau,
1978).

Senators Chaffee and Evans point out that 78 percent
of the national flood insurance claims for 1978 and 1979
were paid to coastal states at a rate three times the
amount collected in premiums:

> Insurance policies in the so-called "velocity
> zones," which are the most hazardous coastal
> areas, cost the U.S. taxpayer about $279 per
> policy, or $14 million annually. (Chaffee and
> Evans, 1981)

The proposed legislation represents an effort to re-
duce federal outlays for the development of barrier islands and for the subsidization of such development by
others in areas clearly vulnerable to natural disasters.

These are just two recent examples of civic leaders
trying to come to grips with public policy decisions
having a scientific or technological component. In the
case of the aerial spraying of California fruit groves,
the decision revolved around the degree of toxicity of
Malathion and the wisdom of spraying the pesticide in a
well-populated area. The second example represented a
situation in which lawmakers have attempted to formulate
a federal economic policy based on our growing knowledge
of environmental phenomena, namely barrier islands, the
effects of urban growth on barrier islands, and the haz-
ards of hurricanes in developed areas.

Colleges and universities could do much more to prepare their graduates for important civic roles--whether
as elected public officials or as citizens--involving
scientific or technological matters. William Wells suggested to the Committee:

It is not so much the details of any particular
science or any particular technology [that are
important in the political arena] as it is the
implications, the impact, the effects of these
areas on other facets of our society. . . . Rad-
ical institutional changes have been under way--
and are under way--with respect to the whole
structure of society as it has evolved in the
western world over the past 300 years. We need a
long view of what these changes are going to mean
for the future of our society. (National Research
Council, 1981c)

David Smith commented from his vantage point at Indiana
University's Department of Religious Studies:

Life in a changing world means that people are
constantly having to learn to cope. Courses
dealing with the human, the value, the moral consequences or implications of scientific change
should be introduced. That many people today feel
religion to be threatened by modern science stands
as a terrible indictment of our educational system, which has irrationally excluded the study of
religion and ethics from its disciplined purview.
There are an increasing number of courses in bio-
medical ethics on college and university campuses;
these represent only the beginning of what can and
should be done. (National Research Council, 1981c)

The public cannot be expected to be knowledgeable
about every facet of the problems that arise from the
extension of science and technology into our culture today. However, to the extent that people are not able to
involve themselves intelligently with the problems at
hand, the solutions will be considered behind closed
doors.

Decisions will be made there, and the rest of us
will be manipulated into agreement. We will be
flattered by being asked our opinion. We will be
presented with carefully selected fragments of
facts and arguments for our consideration. We
will be encouraged to debate, to come together in
block organizations, community meetings, town-
halls, and panels, but will be left unsupplied
with the facts and skills necessary for full self-
determination. (Schwab, 1978)

Even though scientists don't always agree, colleges and universities are obligated to enable non-specialists to appreciate the principles and methods that underlie scientific research and technological development, to know how to seek reliable sources of information, and to reconcile the diversity of opinion and fact in scientific matters that impinge on the well-being of society.

In 1978 the American Association for the Advancement of Science reported that there were nearly 120 programs and more than 900 courses offered by 500 institutions of higher education in the area of "ethics and values in science and technology" (American Association for the Advancement of Science, 1978). These courses treated such areas as the control of science and technology; science and technology's relation to the arts and humanities; the stewardship of natural resources; industry, business, and society; and technology assessment and forecasting. These would certainly seem to represent an opportunity for undergraduate non-specialists who have been introduced to the basic sciences and to the scientific method to learn to deal as effectively as possible with the decisions that await them as citizens and professionals.

College education has the potential and the responsibility to contribute to the preparation of students for civic roles in our scientific and technological society. College faculties should be encouraged to explore ways to awaken undergraduates to the social consequences of scientific research and technological development, as we will discuss in the next chapter.

3

SCIENCE FOR POETS: AN INADEQUATE APPROACH
TO PREPARING FUTURE PROFESSIONAL LEADERS
FOR A WORLD OF SCIENCE AND TECHNOLOGY

The challenge, as defined in the last chapter, is to
make the science curriculum an inviting and meaningful
experience for non-majors. The ultimate goal should be
to attract and hold potential leaders in college science
classes so they can be taught to analyze scientific
problems critically and prepare themselves for a lifetime career of coming to grips professionally and civi-
cally with a world of computers, space exploration, and
the like. The question is: How well are the nation's
colleges and universities doing at meeting this challenge? The answer is: Not well enough.

Clear evidence of this shortcoming came to light when
the Committee, as reported above, held a day-long hearing on March 20, 1981, to solicit views from leading au-
thorities in various professions about the role of
undergraduate science education in the preparation of
future leaders for their respective fields. Witness
after witness testified about the inadequacies of science courses as presently offered to non-science majors.
Leaders in fields ranging across the professional spec-
trum--from politics to law, to journalism, to business,
to public school teaching, to the clergy--all sounded an
alarm about the state of science preparation of non-
specialists. The following excerpts from the March 20
hearing represent the breadth and depth of the concern
expressed by some of the nation's leading professionals:

I believe that most leaders of business and industry are greatly concerned about the "technical illiteracy" of many college graduates and of the
general public. In a recent talk, Edward G. Jefferson, the newly elected chairman of DuPont,
related the following anecdote: "John Kemeny cap-

32

tured the essence of the technical community's
doubts about the body politic. He said that while
he was chairing the presidential commission inves-
tigating Three Mile Island, he had a nightmare. He
dreamed that after minimal debate the House of
Representatives, by a vote of 215 to 197, had re-
pealed Newton's Law of Gravitation. Maybe the
ghost behind that dream was the state legislator--
atypical to be sure--who once urged that the value
of Pi be set at 3.0, so it would be easier to handle in calculations."

>Robert P. Stambaugh
>Director, University Relations
>Union Carbide Corporation

. . . general education science courses are more
frequently a rhetoric of conclusions than an exciting experience that reveals the nature of science. First-year courses are most frequently dues
one must pay, for faculty as well as students, to
get to the excitement of science. When one recalls
that the elementary education majors typically
take only these dreary first-year courses, it is
easy to conclude that they will learn little from
these courses of value to them as elementary
teachers. How well is the U.S. system preparing
elementary teachers in science for their roles as
elementary teachers? In my humble opinion, if we
planned carefully, we could make it worse.

>Hans Andersen
>School of Education
>Indiana University

We are right in the middle of a major revolution
of technology and very little attention is being
given to the implications of that technology, either at the elementary level or at the university
system.

>William G. Wells
>Head, Public Sector Programs
>American Association for the
>Advancement of Science
>(Formerly, Staff, U.S. House
>of Representatives)

In my opinion, judges, lawyers, and legislators
need to know a great deal that seems to lie in the

field of science and which most of them today do
not know. I think it is equally obvious that the
educational system is not providing much teaching
of these matters to anybody but a few specialists
and most of these seem to go on to teach other
teachers.

> Lee Loevinger, Attorney
> Hogan and Hartson
> Washington, D. C.

The present state of college science
education for non-specialists is the result
of an historical evolution that has
benefited the major but left the non-major
largely neglected.

Such testimonials are by no means isolated complaints
from dissident extremists. The Committee found similar
dissatisfaction among a wide variety of responsible and
intelligent observers—including students, educators,
and other professionals. The present disenchantment
stems from the historical evolution in education that
has seen the sciences develop curricular offerings that
are second to none in the world for dedicated students
of science but that leave much to be desired for non-
majors.

Several discernible trends run through the history of
science education in the United States. First, science
has been allotted a role in higher education from the
very inception of colleges in colonial America. The
strength of emphasis has varied, however, from institution to institution and from period to period, depending
upon the availability of qualified scholars and the ac-
cord reached between scientists and members of the faculty and trustees devoted to the humanities, particularly religion. Second, the gradual democratization of
higher education in this country opened up the study of
science to larger and larger populations but, at the
same time, eventually led to a watering down of the science curriculum, for the general student at least.
Third, as the United States has fought to gain and re-

tain world prowess during eras in which international status depended increasingly on the mastery of science and technology, colleges and universities have strengthened their research and teaching for science majors but largely failed to serve non-majors.

As already noted, many prominent early Americans had a good grasp of scientific knowledge, and pioneer colleges made an effort to include science in the curriculum. By 1800 most colleges taught some mathematics and natural philosophy, some taught chemistry, and a few taught natural history (Rudolph, 1977).

The principal force responsible for the inclusion of science in American colleges early in the nineteenth century was the citizenry rather than the growing community of scientists (Ritterbush, 1980b). When the faculty at Amherst College, for example, criticized its instructional program in 1820 for being inadequate, it was on the basis of the fact that the curriculum failed to meet "the wants and demands of an enlightened public" (Guralnick, 1975). Indeed, historian Allan Nevins has equated the "championship of science" in the curriculum of land grant colleges in the mid-nineteenth century with a "demand for greater democracy in education" (Nevins, 1962).

As the corpus of human knowledge expanded, educational leaders such as Harvard's president Charles W. Eliot concluded that it was only reasonable to expect students to master just a part of the curriculum. Students were thus allowed to become "the architects of their own educational development" (Ritterbush, 1980b). The elective system rapidly came to dominate education in the United States; and for a period of time, students could accumulate credits without gaining basic knowledge in primary fields. Distributive requirements were introduced as a means to ensure that students became acquainted with the principal areas of human knowledge, as was intended originally for higher education.

In the course of this evolution, the science curriculum has simply failed to keep up with the demands to educate the non-science major. As science historian Philip Ritterbush notes: "Whereas most students had studied one or two scientific subjects such as physics, geology, or biology for an entire year of each in the 1890s, by 1920 most could satisfy a distributive requirement by studying only the introductory portion of one subject, without following their classmates up a ladder of electives" (Ritterbush, 1980a). Sixty years later, we can safely say the situation has changed little in this regard.

Between 1920 and 1940, at least 30 colleges and universities adopted programs of "general education," nearly all of which included science. The general education courses in science were intended to meet the responsibility of the college to acquaint students with the character and significance of science in the modern world, and their difference from introductory courses was quite clear. Commenting on the introduction of such a course at Haverford College, chemist William E. Cadbury sounded a concern that was to be repeated periodically during the ensuing years and that the Committee has heard many times during its deliberations. Cadbury declared: "Many of us now feel that it is unreasonable to expect a given course to serve simultaneously as general education for some students and as the start of specialized education for others" (McGrath, 1948).

Following World War II, the nation launched a significant effort to strengthen science and technology in every way. Vannevar Bush set the stage at President Franklin D. Roosevelt's request by outlining the form such a national effort might take in the treatise Science: The Endless Frontier (Bush, 1945). By 1950 the National Science Foundation had been established and designated the lead agency in our effort to continue to make new inroads in scientific research and technological development and in science education at all levels (Waterman, 1960).

Recognizing that science affects the life of every contemporary individual, the President's Science Advisory Committee in 1959 concluded that the nation's commitment to the improvement of science education had largely overlooked the education of citizens. In particular, the Committee faulted scientists for failing to provide the kind of teaching material that colleges need for the general student. The President's Committee observed:

Neither the standard course intended for the future professional scientist nor the discursive and frequently fragmentary "survey" course is appropriate. Courses are needed which help a student think his way through and appreciate such great concepts as the origin and evolution of the universe and life, the nature and behavior of energy and matter and radiation, the structure of atoms

and molecules, and the ways in which these and
other scientific concepts and laws are discovered,
evolved, and tested. (President's Science Advisory
Committee, 1959)

At a conference on science in general education at Harvard University earlier in that decade, biologist Paul
Sears observed that introductory subject matter courses
continued to function as a means of selecting out those
students who lacked the aptitude to follow the disciplinary sequence, resulting in an "intolerable neglect"
of future non-scientists. Sears reported to the
conferees:

> I have had colleagues who admit that only 10 to 15
> percent of their beginning students go ahead.
> When I say, "You mean the rest can go to the
> Devil?" they say, "Yes, as far as we are con-
> cerned." (Cohen and Watson, 1952)

Twelve years later physicist Gerald Holton used a
metaphor to illustrate the very same point:

> The classroom usually resembles a training ground
> at the foot of a large mountain that is to be con-
> quered stage by stage by selected students in
> later years. Here, next to the boy who has large,
> high altitude lungs and who was born with climbing
> boots on his feet, there sits by administrative
> decree the eternal lowlander, the stolid farmer,
> the congenital subway rider, the dreaming sailor,
> and even the adventurous deep-sea diver. Silently
> do these listen and move through the mass of tech-
> nical instructions guaranteed to pay off in the
> exhilarating climb to the top--in which, alas,
> they will never take part. (Hoopes, 1963)

Today, just as in the 1960s situation which Gerald
Holton describes, most general students seldom experience the mountain top exhilaration of science. According to evidence the Committee has gathered, non-science
majors are still apt to become bogged down in accelerated introductory courses for pre-meds or be treated to
some watered-down variation of "Science for Poets."
Of course, there are certainly significant exceptions. Some non-majors do have rewarding experiences.
Furthermore, many of the criticisms brought against sci-

ence education, we suspect, apply equally to other disciplines. Our concern here is not with situations where all is well, however, nor is it with the humanities, the social sciences, or the arts. Our charge is to examine college science education for the non-specialist. And we think it is an important commission, for every college graduate is going to have to live in a world where it is difficult, if not impossible, to escape the undesirable consequences of the misuse of science and technology.

In carrying out its charge, the Committee discovered that about 85 percent of the 5.5 million students enrolled in our nation's four-year colleges and universities in 1979 were required to study science beyond what they may have had in high school. This chapter focuses on what we have learned about how much science these students take, what they choose to study, and how well their encounter with science prepares them for dealing with science and technology throughout their professional lives. In the pages that follow, the Committee provides evidence (1) that institutional commitment to the study of science has generally declined, (2) that students are permitted to choose rather freely from a smörgasbord of courses that do not necessarily give them a sound understanding of the basics of science, and (3) that college science education often suffers because of inappropriate classroom materials and inadequate teaching techniques.

Institutional requirements in undergraduate science have declined in the past two decades.

In order to understand the extent to which colleges and universities consider the study of science an important component of undergraduate education, the Committee studied the college requirements and science electives of a sample of 215 post-secondary institutions. The methodology of our survey is described in Appendix A.

We found that the majority of undergraduate four-year colleges and universities today require students to devote only about 7 percent of their total undergraduate course work to the study of the sciences (Table 1). In an institution requiring 125 credit hours for graduation, for example, students will have to devote about 40 credit hours to fulfilling general education requirements, of which about 9 credit hours will be in the sciences. This means that the requirement can be met by taking one full-year course or two half-year courses--

TABLE 1 Proportion of Undergraduate Education in General and in the Natural Sciences at Four-Year Institutions (in percent)

	Carnegie Council Study[a] of Academic Year 1967	NRC Study of Academic Year 1980
General education requirements (mean)	43.1	33.3
Natural science as a proportion of general education	21.0	20.7
Natural science as a proportion of total undergraduate requirements	9.1	6.9

[a] Adapted from Blackburn et al., 1976. That report also included an analysis of general education requirements in 1974. The authors found that general education requirements represented 34 percent of the undergraduate curriculum in 1974, with the natural sciences as a proportion of general education at 18 percent, and the natural sciences as a proportion of total undergraduate requirements at 7 percent.

Sources: Blackburn et. al. in Missions of the College Curriculum, Carnegie Foundation for the Advancement of Teaching, 1977; National Research Council, Survey of College Science Curriculum, 1981.

hardly enough to provide an introduction to the biological and physical sciences and technology.

It is important to keep in mind that the figure of 7 percent represents a national average. Some institutions require more than one full year of science study by their undergraduates, while other institutions have no general education--and therefore no science--requirements at all.

The vocationally oriented undergraduate may think that 40 credit hours is a lot of time to devote to distributive requirements. In actual fact, there has been a substantial erosion of general education requirements over the decades, which in turn has affected the amount of time undergraduates are required to study science. As recently as 1967, students spent about 43 percent of their total undergraduate coursework on general education. Today, general education represents only one third of the total, more than a 20 percent decline since 1967 alone (Carnegie Foundation for the Advancement of Teaching, 1977).

As the fraction of the curriculum devoted to breadth studies has declined, requirements for the study of the natural sciences have also declined by 20 percent--from an average of 9 percent in 1967 to 7 percent in 1980. This typically follows little more than two years of science in high school and little math beyond high school algebra. Fourteen years ago, the same student would have graduated with at least 12 credit hours of science following three or four years of science studies in high school. Despite the continued accelerated advances in scientific research and technological development, non-specialists are actually completing college today with less experience in science than graduates 15 or 20 years ago had. The needs and the trends are clear, and they do not match. It makes no sense for colleges and universities to be requiring less science education at a time when there is an astounding explosion of scientific knowledge and in an age when all of us are touched for better or worse by scientific developments that are often baffling to the uninformed.

TABLE 2 Expansion in the Number of Courses Available to
the Non-Specialist in a Physics Department of a Major
Research University in 1960, 1970, and 1980

1960	1970	1980
1. Elements of Physics, Mechanics, Heat, and Sound	1. Elements of Physics, Mechanics, Heat, and Sound	1. Fundamentals of Physics I
2. Elements of Physics, Magnetism, Electricity, and Optics	2. Elements of Physics, Magnetism, Electricity, and Optics	2. Fundamentals of Physics II
	3. Introduction to Physics	3. Contemporary Physics
	4. General Physics for Science Teachers	4. Physics of Music
		5. Light Perception, Photography, and Visual Phenomenon
		6. Light Perception, Photography, and Visual Phenomenon (laboratory)
		7. Physics in the Modern World
		8. Energy and the Environment
		9. Topics in Contemporary Physics
		10. Basic Concepts in Physics I
		11. Basic Concepts in Physics II

Total Number of Undergraduate Courses

| 36 | 45 | 60 |

Source: National Research Council, Survey of College Science Curriculum, 1981.

As general distributive requirements have
declined, the variety of science courses
across a wide spectrum of special topics
courses has increased.

Following our study of broad field requirements, the
Committee examined the specific types of science courses
available to the undergraduate non-specialist today.
Data were derived from our study of 215 post-secondary
institutions and categorized according to course offerings in five fields: biology, chemistry, physics, mathematics, and computer sciences. We were interested in
estimating the proportion of courses tailored wholly or
in part to the needs of undergraduate non-specialists.
The Committee distinguished between those courses that
represent an introductory encounter with scientific subject matter ("traditional subject matter courses") and
those that offer more advanced treatment of special
topics ("special topics courses"). Examples of each are
given in Appendix B.

Our findings are as follows: While the total number
of general distributive science course requirements has
declined, the variety of courses for non-majors to
choose from in fulfilling those requirements or in selecting electives has proliferated greatly. One need
only survey the college catalogs of the past 20 years to
discern the trend. For example, at one major research
university in the mid-Atlantic region, the department of
physics increased its total course offerings from 36 in
1960 to 60 in 1980 (Table 2). During the same time
span, courses in that department open to non-specialists
jumped from only two to a total of 11 in 1980. These
courses were available to majors and non-majors alike in
1960 but were largely intended for non-specialists as a
special audience in 1980. This growth pattern is con-
sistent with the results of an investigation by the
Chronicle of Higher Education, which found that the number of courses offered by colleges and universities increased by 15 percent between 1979 and 1980 alone
(Magarrell, 1981).

Data from our survey reveal that relatively little of
the teaching effort of science departments is dedicated
primarily to non-specialists, however. If introductory

Proportion of Total Undergraduate Science Courses for Non-Specialists

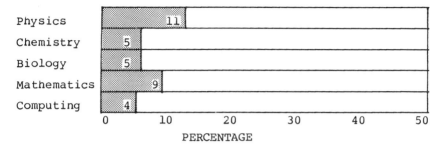

Types of Undergraduate Non-Specialist Science Courses

[a]Traditional subject matter courses for non-science majors such as business or education (specific) and courses that do not target any particular non-specialist group (general).
[b]Special subject matter courses for non-science majors that attempt to teach science within an integrated or interdisciplinary framework using thematic, historical, social, or popular approaches. Many have removed all mathematics requirements.

Source: National Research Council, Survey of College Science Curriculum, 1981.

FIGURE 1 Proportion of undergraduate science courses for non-specialists by field and by type of course

TABLE 3 Proportion of Institutions Offering at Least One Course for Non-Specialists by Field, Course Content, Institutional Type and Control

	Number of Institutions (N)	Field (percentage of institutions)				
		Physics	Chemistry	Biology	Mathematics	Computing[a]
Traditional Subject Matter Courses						
Courses for non-science majors						
Research university	47	81	26	45	79	32
Doctoral university	32	69	31	53	78	38
Comprehensive university/college	110	56	43	56	86	31
Liberal arts college	26	35	23	19	62	0
TOTAL	215	61	35	48	80	31
Public university/college	126	71	37	57	89	29
Private university/college	89	46	30	36	67	35
HPBC[b]	16	31	31	44	69	44

44

Special Subject
Matter Courses

Courses for non-science majors						
Research university	47	75	40	68	40	26
Doctoral university	32	69	47	75	31	25
Comprehensive university/college	110	59	38	57	45	28
Liberal arts college	26	39	23	39	27	0
TOTAL	215	61	37	60	36	20
Public university/college	126	75	40	67	44	32
Private university/college	89	43	34	51	34	10
HPBC[b]	16	50	6	19	25	33

[a]Computer science departments or divisions have not been established in all post-secondary institutions. Of those surveyed, 39 research universities, 24 doctoral universities, 83 comprehensive universities/colleges, 7 liberal arts colleges, and 9 historically or predominantly black institutions made such distinctions.

[b]Historically or predominantly black colleges.

Source: National Research Council, Survey of College Science Curriculum, 1981.

subject matter courses taught to science and non-science majors alike are excluded, the average proportion of undergraduate courses being offered for non-specialists has this breakdown by discipline: chemistry and biology, 5 percent; computer science, 4 percent; mathematics, 9 percent; and physics, 11 percent.

Most of the teaching effort of science faculty members for non-specialists focuses on special topics courses (Figure 1)--many of which were designed as a response to the outcries of the 1960s and early 1970s for relevancy in higher education. These courses address such topics as ecology and human society, the physics of sound, and environmental chemistry. More than half the courses for non-specialists in physics, chemistry, and biology assessed in our survey were of this genre. (An exception to the pattern is evident in computer science and mathematics, where the primary emphasis is on offering general introductory courses for special audiences such as business majors or education majors.)

While advances have been made by the science community in the development of courses for undergraduate non-specialists, the effort has not been uniform when analyzed by type of institution (Table 3). Physics departments in research universities and in public postsecondary institutions are more likely to offer separate introductory subject matter courses for non-scientists than are physics departments in liberal arts colleges, for example. This same difference is evident in the fields of biology, mathematics, and computer sciences. The Committee also notes that, with the exception of mathematics, fewer than half the historically or predominantly black colleges have developed separate introductory subject matter courses for non-specialists in the science fields surveyed (Table 3).

While our data suggest that greater emphasis has been given to special topics courses for non-specialists than to the development of introductory courses, the effort again has not been uniform across colleges and universities. Research universities and public institutions are more likely to offer special topics courses in every category of science studied (Table 3).

Another finding from our study is that courses for undergraduate non-specialists in the computer sciences are almost nonexistent at liberal arts colleges. It is unclear from our survey, however, whether the absence of such courses at liberal arts colleges reflects fewer resources to provide such courses. Further study of this finding is obviously needed.

In summary then, undergraduate non-specialists enrolled in research universities seem to enjoy a greater degree of choice when selecting courses to fulfill their undergraduate science requirements. Non-science students in liberal arts colleges, in contrast, generally have a narrower range of courses to choose from, although the factors contributing to this difference are not known and would need to be analyzed before firm conclusions could be drawn about the finding.

The findings from a recent study of our nation's two-year colleges have yielded results parallel in many ways to the outcome of our own investigation. A series of monographs issued under the direction of Arthur M. Cohen, University of California, Los Angeles, reported little evidence of science courses appropriate for non-majors in two-year colleges today (Beckwith, 1980; Edwards, 1980). Mooney, for example, commenting on the availability of special courses for non-concentrators in physics, noted that the study did not support the observation of some educators that two-year institutions are providing leadership in the development of special courses for non-majors (Mooney, 1980a). These colleges contribute to the education of future civic leaders, and we believe that greater effort should be made in such institutions to ensure that appropriate education in science is made available to non-specialists.

It is one thing to know how many courses are made available to undergraduate non-specialists in the sciences. It is another to know what courses students actually elect to take. The Committee did not have the resources to examine the course-taking behavior of undergraduate non-specialists in detail. Instead, we conducted a limited set of interviews with about 20 science faculty members in a sample drawn at random from the set of 215 institutions surveyed. They were asked about the enrollment, format, and content of their courses. This information supplemented our perceptions of the situation based on the professional experience of Committee members and opinions expressed at our various hearings.

Our findings suggest that interest in the many special topics courses for undergraduate non-specialists available today has probably peaked. Enrollments in topics courses in physics--such as "Physics for Poets," "The Physics of Acoustics and Music," and "Physics and Society"--hover at an average total enrollment per semester of 10 to 20 students, according to the instructors surveyed. Furthermore, enrollments seem to be shifting according to changing student attitudes on what actually is relevant. One science educator we spoke with, for example, who has received significant amounts of federal support for course development over the years, told the Committee that "The Physics of Music" is no longer of interest to undergraduates, but the more current topic "The Physics of the Environment" is. Textbook publishers confirm this observation.

Our findings also indicate that there is a growing body of critics, both professional and non-professional, who think that colleges and universities may have reacted too hastily in meeting ephemeral student pressures to create topics courses. These critics charge that in some--perhaps many--cases, student demands for relevancy may have resulted in topics courses that are vacuous or artificial or both.

Whatever the particular merits of such criticisms may be, certain general trends and conclusions about the proliferation of course offerings do seem in order. First, the proliferation of courses has taken place in a rather haphazard manner without any long-range or coordinated planning as to how such offerings fit into an overall plan of liberal education. The result is that students are treated to a smörgasbord of course offerings that many times stress relevancy over mastery of basic scientific principles. Topics courses, when taught at their best, provide a basic grounding in the fundamentals of some scientific discipline so that students are prepared to argue opinions that are based upon facts. Second, the wide array of courses can often prove puzzling and confusing to students because they lack the necessary counseling to line them up with the proper courses in light of their particular preparation and educational needs. The unfortunate result is that some students are able to hopscotch through their so-called scientific education without any apparent direction or coherence to their learning. None of these circumstances do much either to prepare future leaders with a critical facility to attack scientific questions or to give them the knowledge and experience they need to handle the technological demands of their professions.

> Basic introductory science courses often-
> times fail to reach their full potential
> because of ill-prepared students and
> inadequate teaching.

Other than special topics courses, the main and certainly predominant route for non-specialists to gain
scientific knowledge is through general beginning
courses, such as "Introduction to Biology" or "Organic
Chemistry" or "Basic Physics." Yet such courses offer
no panacea. Say the phrase "introductory course in science" to most students or former students and, according
to our finding, they are likely to respond: "huge
classes," "weed-out course," "sleep," "dull," "boring,"
or "useless." The effect on the non-science group--the
captive audience--is especially unfortunate. One student, majoring in economics, put it this way:

> I had a bad experience in chemistry. It was the
> worst course I ever had. The course emphasized
> memorization over learning the theory of chemis-
> try. . . . Part of it had to do with the class
> being so large. . . . I guess there were 1,000
> students. The computer was used to grade tests and
> check lab results. It was all very impersonal, but
> I guess the large number of students forces them
> to use that system. (National Research Council,
> 1981a)

Part of the difficulty lies with the students. They
enter such courses ill-prepared to understand the concepts being treated or to undertake the study of science
at the collegiate level.

At one time, the science education provided by our
colleges and universities was better integrated with the
science taught in our high schools. Therefore, students
were more likely to emerge from college having had sufficient exposure to science at either or both levels of
education (Rudolph, 1977; National Research Council,
1980). As colleges relaxed their entrance criteria and
high schools modified their requirements for graduation,
less emphasis was placed on preparation in science for
those not majoring in science-related areas. The result

of the disengagement of this "vertical integration" is
twofold. In the first instance, students are more
likely to experience feelings of inadequacy when con-
fronting science in college if their high school science
preparation has not been appropriate for further education. In the second instance, college science faculty
find it more difficult to strike a proper balance between the science being presented and the ability of the
students to handle the information.

In a hearing convened by the Committee in December
1980 to discuss undergraduate science instruction for
non-specialists, Arnold Arons, professor of physics at
the University of Washington, described the mismatch
that frequently occurs now between the "curricular
materials and the minds of the students that are sup-
posed to receive the materials." He explained:

> The fact that emerges is that in our science
> courses at colleges and universities, we take
> material that requires abstract logical reasoning
> of various kinds, and--without any attention paid
> to the students--we throw the material at them as
> though they were completely ready for it. . . .
> Much of what we are doing at the college and uni-
> versity level drives our students into blind mem-
> orization instead of into comprehension and under-
> standing. . . . If we want to reach the non-
> specialists, it seems to me that we have got to
> give them time to make mistakes, to retrace their
> steps, without being punished for being "wrong."
> (National Research Council, 1981b)

Clearly, the readiness of the student to receive scientific information and to use scientific concepts should
be a critical element in designing introductory science
courses for specialists and non-specialists alike. The
evidence is that this fact is too seldom recognized.

Secondary schools should not bear the full indictment
for the ailments of introductory college science
courses, however. The successful classroom is as much,
or more, dependent upon good teaching as it is on ready
and willing students. All too often, according to both
professors and students who appeared before the Committee, those enrolled in introductory college science
courses do not experience teaching at its best. Sometimes professors rush through lectures in order to get
back to the more exciting atmosphere of their research

laboratories. Oftentimes the discussion sections and laboratory are simply turned over to graduate teaching assistants.

In an attempt to improve the situation, faculty in the sciences have taken an interest over the years in developing introductory subject matter courses with conventional content but adapted to the needs of the non-scientist. The results of these efforts appear mixed. Of the faculty we interviewed, those who have developed this type of course in the biological sciences seem to be enjoying the most success, at least when measured by size of enrollment. Traditional subject matter courses for non-scientists in physics and in chemistry will usually have fewer students enrolled, on average, than comparable biology courses for non-specialists. In our analysis, the biology enrollments out-numbered those in physics and chemistry 10 students to 1.

Interpretation of the apparent success by biology instructors in developing courses of interest to non-specialists is confounded by a general factor of student preference for biology. We believe that non-specialists are more likely to elect a biology course in college because they are familiar with the subject matter from high school, because the courses require little familiarity with mathematics, and because the students perceive the topic to be relevant to their personal health interests. Advisors to liberal arts students told us that students like biology because they can use the information in their own lives.

We do not mean to suggest that there are not a number of educators in the non-biological sciences who have been successful in developing valuable and popular courses for undergraduate non-specialists. There are. Our impression is, however, that in a system which allows students to choose among the natural sciences in fulfilling the undergraduate distributive requirements, undergraduate non-specialists will be more inclined to select biology--and possibly the earth sciences--rather than physics, chemistry, or other more quantitatively-oriented college science fields. For whatever reasons, undergraduate non-specialists have generally narrowed the range of science options to those fields within which they feel they can comfortably operate. We believe this is an unfortunate turn of events because "breadth in science" can be _every_ _bit_ as important as "breadth in general education" in an age when advances continue to be made in every field of scientific research.

Despite these worthy efforts to design introductory classes especially for non-specialists, serious teaching problems still remain. Our survey findings suggest that teaching aids--such as lecture demonstrations and films --and science laboratory experiences are declining in use. As one physics professor put it, they had to eliminate labs for non-science majors because the costs of operating them were too high given their department budget. This economizing may be necessary, but it is regrettable.

For many non-specialists, the concepts, principles, or vocabulary of a science are in danger of remaining meaningless in an introductory course unless some provision is made to provide students with a firsthand experience with phenomena. We believe this "hands-on" experience is crucial to the understanding of the science. The use of many demonstrations, models, and simple laboratory experiments adds reality to the pursuit of scientific knowledge. The excitement of scientific discovery can be transmitted all the more effectively and meaningfully if the student has the opportunity to experience the subject of study through his or her own senses and with instruments that are extensions of those senses.

In the final analysis, good teachers--and good teachers alone--are the key to solving not only these classroom difficulties but all of the shortcomings delineated in this chapter. Bright, knowledgeable, and inspired teachers who are truly dedicated will somehow find ways to overcome the institutional and curricular barriers to preparing non-specialists for leadership roles in an era of scientific and technological advancement.

4

ELIMINATING BARRIERS TO AN
APPROPRIATE UNDERGRADUATE EXPERIENCE IN SCIENCE

The key to eliminating the barriers that prevent colleges and universities from reaching their full potential in teaching non-specialists science is human ingenuity and dedication. To put it succinctly, we must attract highly motivated and talented teachers to meet the challenge of educating non-majors about science and then provide those teachers the means of fulfilling their calling.

This requires that a number of conditions be met. First, there must be an appealing incentive for taking on and achieving the task. Second, these quality teachers must be guaranteed adequate time with students to fulfill their curricular goals. Third, there must be an adequate vehicle in the form of courses for executing the teaching mission. Fourth, the faculty should be provided a forum for brain-storming curricular ideas with science colleagues as well as with leaders in the professions. Fifth, those professors taking on the task must have appropriate teaching tools, such as audio-visual aids and laboratory material. Sixth, there needs to be a national support system to help provide leadership and disseminate model course materials and innovative ideas about teaching non-majors.

Some of the changes required to meet these provisions are attitudinal. Others require commitments of resources: free time and in some cases funding. The financial requirements are not necessarily great, however. In many cases, the end goals can be accomplished by redirecting existing fiscal resources. In other instances, channels and operations already in existence can be tapped. We turn now to specific recommendations for achieving the end results.

54

RECOMMENDATION 1

The Committee urges colleges and universities to find
new and additional ways to identify and reward high-
quality teaching of science courses for non-specialists. Prizes, sabbaticals, and increased consideration
of teaching contributions when tenure and salary decisions are being made should all be a part of a planned
incentive program by higher education, working in concert with governmental bodies and the private sector.

On the eve of his departure from the White House in
1961, Dwight D. Eisenhower accurately forewarned that
changes lay ahead in the nature of universities and pre-
dicted that university scientists would become more concerned about how to compete successfully for support of
increasingly specialized research:

> Today, the solitary inventor, tinkering in his
> workshop, has been overshadowed by task forces of
> scientists in laboratories and testing fields. In
> the same fashion, the free university, histori-
> cally the fountainhead of free ideas and scien-
> tific discovery, has experienced a revolution in
> the conduct of research. Partly because of the
> huge costs involved, a government contract becomes
> virtually a substitute for intellectual curiosity.
> For every old blackboard there are now hundreds of
> new electronic computers.
>
> The prospect of domination of the nation's schol-
> ars by federal employment, project allocations,
> and the power of money is ever present--and is
> gravely to be regarded. (Eisenhower, 1961)

Scientists employed in the academic setting indeed
have been forced by the tide of events to become more
specialized and preoccupied with the competition for
research dollars. As a result, they spend more time on
research and less on teaching. Several studies have
shown that science faculty devote slightly more than one
fourth of their total work time on average to teaching,
although this figure varies from field to field and from
institution to institution (National Research Council,
1980).
Part of the explanation for this trend, no doubt,
stems from the fact that teaching is generally not

highly esteemed by the public. Public opinion polls
show that, among the white-collar professions, medicine
and law rank highest in public esteem, followed by several other professions. Teaching falls at the bottom of
the list (Isaacson, 1971). One upshot of all this is
that many college science faculty see themselves as research scientists first and as classroom teachers second. Teaching all too often is regarded as a duty of
employment, although many outstanding investigators are
also known to be outstanding teachers.

The Committee recognizes the important part played by
science faculty in our national research effort. We
believe, however, that emphasis should also be given to
elevating the role of teaching in research-oriented departments. The rewards system within this setting, and
to some extent throughout the science profession and
society at large, does not do enough at the present time
to enhance the status of science teaching of non-specialists in post-secondary institutions. Promotions
within the science departments and decisions regarding
tenure, particularly within research universities, need
to be based more strongly on good teaching as well as
good research. Equally important, society--through governmental bodies and the private sector--needs to help
build a better reward system for teaching.

Quality teaching can also be encouraged through financial incentives introduced by the states to stimulate
innovations at public institutions. California, for
example, has introduced a program that permits the University of California system to award grants of about
$5,000 to faculty members to improve the curriculum.
These funds allow faculty to develop new materials for
the classroom, hire teaching assistants, or acquire
slides for teaching aids.

Scientific societies are in an especially favorable
position to play a vital role in raising the status of
the faculty instructor within the ranks of the profession. Awards for excellence in college science teaching
for non-specialists cost little to the professional society and are very effective. Inviting innovators in
college science education to address society members at
annual or regional meetings, especially when portions of
the program are set aside for teaching symposia, is also
effective and allows more visibility for those engaged
in advancing undergraduate science education. Some of
the larger scientific disciplines already have effective
societies, associations, or other units dedicated to the
improvement of teaching. More need to follow suit.

Business and industry should do much more to encourage good teaching than they do now. By making funds available to colleges and universities, businesses could help establish named awards for excellence in the teaching of science to undergraduate non-specialists. These awards could involve cash prizes or grants to encourage further innovations in teaching.

Internship awards also could be designed to bring teaching faculty into the business or industrial setting for brief periods of time to help teachers become more familiar with the professional areas served by their undergraduate science courses. For example, a biologist who has been recognized for his contribution in teaching a course to undergraduate non-specialists on the ethical implications of genetic engineering might spend two or three weeks in an industrial research laboratory. Thus he could become familiar with research advances, state-of-the-art considerations, legal questions, and--through formal or informal discussions with industry-based peers--new aspects of the ethical questions posed by this type of scientific advance.

Help from the federal government is needed in providing such incentives. With the assistance of the National Science Foundation, instructors should be recognized through White House awards, perhaps called the President's College Science Teaching Awards. The prestige brought by this type of national acclaim--the details of which are provided in the next chapter--would elevate good teaching in the public's perception while strengthening the perceived value of college teaching within the science community.

Clearly, then, there are ways--some of them quite inexpensive--to increase the rewards for good college teaching and consequently to raise the esteem of academic instruction of non-specialists with the institution, in the field, and among the public.

RECOMMENDATION 2

In light of declining science requirements over the past two decades, the Committee encourages colleges and universities that have lowered their science demands for graduation to reverse direction and raise their requirements. We believe that a total of no less than two one-year courses selected from the biological and physical sciences and mathematics should be required of non-specialists for the baccalaureate degree.

No matter how dedicated and qualified teachers are,
they cannot prepare non-specialists in the sciences unless they have adequate time to impart knowledge. The
135-contact-hour national average devoted to science
brought to light by our survey is simply not enough. We
are concerned that the requirements in some, but
certainly not all, colleges and universities are subminimal. The 9-credit-hours average required in institutions is only enough to turn out students who are
barely functionally literate in science and technology.

We believe colleges and universities are obligated to
help each student acquire some measure of knowledge of
each of the main fields of human inquiry, including the
study of science. Our concern is that science be accorded once again a full and appropriate role in the
undergraduate curriculum. This concern transcends the
traditional view that liberal learning contributes to
the refinement of the individual (Eliot, 1915; Snedden,
1931; Rudolph, 1977). While the cultivation of the individual certainly represents an important and laudable
goal of college education, we believe the study of the
sciences by undergraduate non-specialists is important
because it bears directly on the capacity of those individuals to operate effectively in an increasingly scientific and technological society. Lawyers, journalists,
business people, and the clergy alike often look to a
liberal arts education to provide them with the broad
knowledge base that they will need to fulfill their
ultimate responsibilities as citizens and leaders in
their professions (National Research Council, 1981c).
This is the breadth in learning that a carefully planned,
well-executed program of liberal arts education can provide.

Christine Harris of the Consortium for Minority Journalism, for example, in anticipation of her testimony
before the Committee on March 20, 1981, interviewed a
number of black journalism educators about how much and
what type of science education journalism students
need. She noted that all the persons she interviewed
agreed that "there are just too many science-related
issues journalists must cover today" for science to be
neglected at the undergraduate level, and that there was
"general agreement that the best education a journalist
can have is a solid and broad liberal arts education
that includes science and math" (National Research Council, 1981c). Malcolm Mallette, director of development
for the American Press Institute, pushed the point even

further. He emphasized the role of liberal arts education in preparing newspaper reporters to become "good generalists." He put it this way:

> Any journalist needs a grounding in history, political science, economics and sociology, among other subjects. That is why journalism courses are limited to 25 percent of the undergraduate curriculum. But with all those needs, I would still hope that all journalism students would take undergraduate courses in math, chemistry, and physics. They will then be better prepared as generalists and in a position to take specialty training in science if they wish something additional after the bachelor's degree. (National Research Council, 1981c)

Similar comments regarding the importance of liberal arts learning and the role of science education in that context were provided by representatives from the fields of law, business, and religion among others. Harold Green, for example, told the Committee that as a graduate of the University of Chicago during the Robert Maynard Hutchins era, he believes that:

> . . . a college education should provide every student--whatever the discipline, profession, or vocation to which he or she is bound--with a broad, general education that consists of at least a general survey course in the biological sciences and the physical sciences, and of course, in the social sciences and humanities as well. (National Research Council, 1981c)

Jerrier A. Haddad, vice president for technical personnel for the IBM Corporation, told the Committee that college could play an important role in providing individuals who will work in management positions one day a knowledge of scientists and of engineers "with respect to their goals, their ambitions, their rewards, their frustrations, their methods, their practices, their lines of reasoning." Haddad observed that the undergraduate system of education as it is presently designed does not attack "this set of elements" in his view (National Research Council, 1981c).

College educators in this nation need to think through their science offerings for non-science majors

in light of the demands of contemporary society and the
professions. Campus by campus, educators need to come
forth with a curricular plan that ensures that non-specialists will leave college with an understanding and a
command of science necessary to survive and prosper in
the modern world. We believe that requires a minimum of
two years of science study. Wherever possible, the federal government should assume a facilitating role in
this process by providing data, coordinating planning,
and communicating results without infringing on the traditional rights of higher education to control its own
destiny.

RECOMMENDATION 3

The Committee recommends that college science faculty
restructure introductory subject matter courses and redesign special topics courses to meet the changing educational needs of undergraduate non-specialists. We believe the federal government, together with the private
sector, should make financial resources and awards
available to realize this goal.

In the preceding chapter the Committee reported that
courses for non-specialists have proliferated but that
the content of many fails to meet the needs of non-specialists. Having made this finding midway in its study,
the Committee considered whether it should describe in
some detail what this content should be. What were the
basic principles that surely must be included? What
were the interest-exciting applications that might be
made? How long should such courses be, how should the
time be distributed over the various topics, and when
should they appear in the undergraduate curriculum? The
matter was debated at some length, but in the end most
members of the Committee were disinclined to engage in
such a venture. One reason was the lack of time and resources to do this well. Among other considerations was
the fact that major fields of science were represented
by at most a single member of the Committee. A more important conclusion was that, even if resources and personnel were available, it would still not be an appropriate task for this Committee.

Detailed course-content design and curriculum development, we believe, are the responsibility of science
faculty members working on their own campuses, in insti-

tutional groups, or through their professional associations. The problem we are addressing needs to be brought to their attention, and they need to be given the encouragement and the resources to solve it. But they must do the job. They are the ones who know their fields, their students, and their institutions.

The Committee decided rather to try to describe some general strategies for making progress and to suggest the conditions that would make these strategies successful. What follows in this section of our report is of that nature and not a detailed treatment of specific course content.

We believe that special topics courses have an important role to play for students who already have a solid grounding in science. We suspect, however, that many of the special topics courses being offered have simply outlived their utility and relevance. The many new directions science and technology have taken us in the past decade alone, and the further new directions on the horizon, suggest that special topics courses need to be revamped.

Furthermore, better counseling is needed to help students find their way into the right topics course in light of their science background and needs. We strongly urge science departments to reach out and work closely with the professional disciplines especially in helping build a network of academic counselors who are interested enough and informed enough to guide non-specialist students in this type of coordinated counseling.

We believe special topics courses would benefit from stronger interdisciplinary ties of still another sort. Science faculty also need to orient themselves more directly to the concerns of the non-science community in designing the courses. Cooperation with other disciplines in all likelihood would result in the development of special topics courses that better permit students to consider how a science field interfaces with the professional considerations of a non-science field.

There also should be adequate opportunities for interested non-science faculty actually to help teach topics courses. Several offerings have been developed in recent years by interdisciplinary teams. These ventures have encouraged students from many different disciplines to consider such issues as the implications of biotechnology for health care; a literary perspective on scientific ethics; a social history of the impact of machines on American institutions; and the ethical, social, and legal control of broadcast technology (American Association for the Advancement of Science, 1978; National Research Council, 1981). A side benefit from such cooperative efforts, no doubt, would be exactly the kind of knowledge exchange necessary to establish the type of counseling called for above.

We also urge college science faculty to place greater emphasis in the near term on the development of introductory science courses that better meet the needs and abilities of undergraduate non-specialists today. We suggest, for example, that science faculty recognize the important role they can play in designing courses that help alleviate those fears about science that plague students.

There is some evidence that undergraduate science faculty help overcome science anxiety by developing courses that stress the basics of science but are designed for specific fields--science for the business major, science for religion majors, science for the journalism major. While there certainly seems to be a role for basic science courses designed to meet the science education needs of the various professions, the continued development of such courses should be approached with some caution. Too much emphasis by the science community on this type of course might restrict the breadth of the educational experience. We believe that the scientific knowledge that future non-science professionals require can be provided in a general purpose science course, especially if care is taken to provide for appropriate applications.

College science faculty, together with their local administrators, will have to determine what their resources will permit in improving introductory science for non-specialists. Smaller science departments that cannot educate non-specialists separately can review and seek modifications to serve both the science major and the non-specialist. It may be possible to provide greater opportunities for non-specialists to explore science at a level they can handle through question-and-answer sessions in the lecture, or through specially designed discussion sections. Discussion sections, for example, might focus on the field of business, or education, or journalism.

We have noted with alarm the trend to eliminate laboratory experience from basic courses for non-majors. There is no reason to equate "hands-on" experience with costly laboratory equipment and increasingly hard-to-

find materials. There are many ways to provide undergraduate non-specialists with firsthand experience with
phenomena without excessive cost to the student or to
the department. Several faculty we interviewed have
adapted readily available kits sold at hobby shops--such
as those that employ pieces not unlike tinkertoys--to
build such things as models of complex molecules. Other
faculty have figured out how to use pieces of kitchen
equipment and other handy devices to demonstrate physical principles. We are aware of courses where the instructors use field trips to the local surroundings in
Utah to study the flora and fauna, aligning the content
of the fall and spring classes to the corresponding veg-
etative and life cycles. Perhaps nothing stands out
more persuasively as an example of a simple, low-cost
demonstration than the piece of pliable cardboard used
by Professor Carl Sagan in his television series Cosmos
to illustrate how ancient Greeks deduced the curved
shape of the earth and its circumference, using differences in the length of shadows at different points along
the surface of the globe.

We believe it is possible to extend the opportunities
for firsthand experience with phenomena in a low-cost
fashion because we have seen how successfully it can be
done.

In our conversations with faculty, we learned that
some science educators are interested not only in providing non-specialists with stimulating experiences with
scientific phenomena, but also in providing them with
experiences that are relevant to their professional interests. Faculty may wish to give special consideration
to the development of upper-level college courses in
science and technology for non-science students who have
made a commitment to a career. Such courses would be
similar to those provided for future elementary school
teachers (see Chapter 2). Such courses would allow undergraduate non-specialists who have demonstrated some
mastery of the basic sciences to explore ways in which
science and technology serve as tools in their profession, or to sharpen their understanding of specific
areas of science that they may one day have to communi-
cate to others. For example, individuals who will work
one day as law enforcement officers, as lawyers, or as
medical writers may be interested in studying the forensic sciences, including what has come to be called
forensic chemistry.

According to several faculty members we interviewed,

there is a real dearth of laboratory guides, demonstrations, or other visual aids that would help college science teachers devise experiences with phenomena that relate the sciences to non-science professions. Most faculty apparently fall back on their own intuitive resources and creativity to extend discussion or to formulate demonstrations that are meaningful to the student who will work one day as a non-science professional. This represents an area in which innovators in science education should be encouraged through external support to develop new materials for use by science faculty.

A cautionary note is in order. We do not want to pretend that the process of getting innovative ideas into production and the products into classroom use is clearly understood, easily implemented, or always in need of stimulation. Many individuals throughout academia turn out excellent educational materials year after year. These are subjected to the market test by colleagues and commercial procedures.

We believe that the production of new courses by large-scale projects in the style of the 1960s and early 1970s should be approached with care. Not all of those early efforts were successful. There may be a need for major course-content projects when the educational approach in an entire discipline needs a complete rethinking, when the needs of special areas are not being well served, or when quality of educational materials has been slipping.

The need for much individual experimentation with new educational approaches seems clear. As we noted in Chapter 1, pluralism is the hallmark of American education. We believe that if there are many individual attempts to improve undergraduate science education of non-specialists, some excellent things will emerge, and there will also be a wider range of choices available to the teacher.

RECOMMENDATION 4

The Committee calls upon colleges and universities to provide a forum for scientists and non-science professionals to explore together new directions in science education for non-specialists. Through regularly scheduled faculty meetings, seminars, or retreats, faculty should be encouraged to develop science experiences appropriate to the educational needs of undergraduate non-

specialists. These efforts should be guided by regular
consultation with leaders in the professions.

 The college science community has come part way
toward the goal of providing courses suitable to the
needs of the undergraduate non-specialist. While such
efforts may not be widespread, there have been significant success stories that need to be sustained.
Furthermore, we need to consider how to extend these
somewhat isolated successful ventures across higher
education in general. Where do we go from here?
 The Committee believes that the quality of undergraduate science education for non-specialists cannot be the
concern of the scientific community alone. The final
authority, however, for deciding course content clearly
should be the decision of science experts. College faculty in other areas, who are responsible for setting the
recommended program of study for non-specialists, must
identify and make a clear commitment to the role of science and mathematics in professional training and the
liberal arts experience. In doing this, they will nat-
urally be somewhat dependent upon the professional community. This means that science faculties should consult
with non-science colleagues--in law, journalism, business administration, precollege education, and the other
professional fields discussed in this report. Such
cooperative curriculum planning should strive to ensure
that the content and presentation of courses satisfy the
educational needs and requirements of undergraduate non-
specialists. If the non-science professional community
can be enlisted in educational planning, there is a
strong likelihood professional leaders will recognize
the potential contribution of science to the education
of non-specialists. In the end, undergraduates very
likely will be encouraged to acquire the appropriate
competence in science.
 The Committee has in fact found among non-science
professionals a great interest in the science component
of the education of people in their fields--an interest
that can form the basis of effective cooperation.
 Colleges and universities are often so large that
faculty--regardless of field--do not know what is going
on in courses being taught down the hall much less
across the campus. Deans of arts and sciences, deans of
faculty, and vice presidents for academic affairs need
to play a more forceful role in bringing representatives
of these teaching faculties together.

It is conceivable that businesses and other sources
of private funding might be interested in sponsoring
seminars or retreats for faculty to find out what students from non-science departments are studying and how
science education might be more responsive to their educational needs.

Science departments should also play a more active
role in reaching out to the non-science community for
ideas about ways to improve undergraduate science
courses for non-specialists. In this context, state
governments might consider providing institutions with
such funds as are necessary to bring faculty together
under the aegis of the various science departments to
discuss new directions in college science education for
the non-specialist within institutions. It is apparent
that the changes in undergraduate science education
needed today can proceed most effectively after some
agreement is reached between and among fields as to what
is needed and what the goals for change ought to be.

RECOMMENDATION 5

The Committee encourages colleges and universities to
extend the use of non-traditional instructional media in
teaching science to non-majors in new and possibly more
exciting ways. Special attention should be given to the
educational potential of mini- and microcomputers and
such public broadcasting ventures as the Annenberg
project.

To attain quality teaching of science to non-majors,
instructors will need to perfect the tools of teaching.
Other than using traditional slides and viewgraphs--and
occasionally lecture demonstrations--most faculty members are possibly not very inventive, and certainly not
very active, in employing the many devices available
today to make college science classes interesting and
lively and to extend the learning experience beyond the
classroom. The literature on the use of such devices is
large and readily available, and the devices themselves
are often within easy reach of most teachers and should
be a regular part of instruction. Professors also ought
to make use of other teaching materials including inex-
pensive supplies and even housewares. In addition, more
recently developed learning aids--computers, broadcast
and cable television, and videodiscs--should be used

more widely and effectively in college instruction than they are at present.

Students who have had experience with interactive computer-aided education generally turn out to be enthusiastic supporters of this approach to learning. One example of what can be done with the computer may be found in the PLATO system.

The PLATO [Programmed Logic for Automatic Teaching Operation] system was developed at the University of Illinois beginning in 1960 and today includes approximately 1,200 terminals scattered across the United States. Users have access to about 16,000 hours of instructional material in more than 200 subject areas. It is estimated that PLATO has the potential of reaching more than 70 million students at all age levels at present.

Even more promising are the advances in microelectronic technology that will make it possible for students in the near future to have access to microcomputers for many diverse educational purposes. For example, commercial educational firms such as Control Data Corporation are developing scientific and other programs that can be used on personal computers. Integrated components probably will soon make it possible to bring together voice, image, and data that can be manipulated at the command of the user (Carpenter, 1980).

In addition to the use of computers in undergraduate education, closed-circuit television and public television have a great deal to offer undergraduate science education. According to a study conducted by the National Institute of Education, telecourses have enabled older students, women, and those who are employed to enjoy undergraduate education (National Institute of Education, 1979).

The recent donation of $150 million to public television by Walter Annenberg should also provide educators with a new opportunity to extend the use of televised instruction (Feinberg, 1981). Annenberg's gift to the Corporation for Public Broadcasting represents the first major national effort in the United States to produce college-level courses on television. The Corporation has indicated that panels of scholars from across the nation will assist in devising courses to be offered through existing colleges for credit. We strongly urge that courses for undergraduate non-specialists be included in the project.

Finally, numerous instructional technologies have the

potential to enrich the undergraduate science experience. Videocassettes and videodiscs, slide-tape programs, multimedia presentations, and audiotapes are just a few examples of media available to the science educator. As a matter of fact, many non-specialists have already found that these teaching devices are being used in classes taught by instructors in their major field. Science educators must increase their use of instructional technology in courses for non-specialists, if more of those undergraduates are to be attracted and given exciting experiences.

Clearly, the technologies are available. The primary challenge now is to use these educational innovations appropriately in meeting the science education needs of the non-specialist.

RECOMMENDATION 6

In light of the experience of the college science commissions in the 1960s, the Committee recommends that all professional societies provide more leadership in educational innovation and propagate information widely about new directions in science education for non-specialists. To the extent they require financial assistance, the federal government and the private sector should supplement funding.

Science faculty often labor in isolation to bring new ideas into their undergraduate courses, some by decision, others by circumstance. Part of the problem is the failure of the institutional system to support the work of potential innovators. Another part has to do with the lack of information available to some instructors about innovative approaches to teaching science to the non-specialist. Information about existing science courses for non-specialists needs to be propagated more widely to give science faculty interested in doing more for their undergraduates a chance to see what others are doing.

How do college science faculty find out what their colleagues are doing in the way of new approaches to teaching? Judging from our interviews with teachers, it varies enormously. A few have established informal ties with colleagues in other colleges and universities. These colleagues critique each other's approaches to teaching and suggest ways to improve instruction.

Others have joined scientific associations devoted wholly or in part to science teaching, such as the National Science Teachers Association, the American Association of Physics Teachers, the Division of Chemical Education of the American Chemical Society, and the Mathematical Association of America. Members follow developments in teaching science through journals, meetings, and association newsletters. These associations also occasionally devote sessions at annual meetings to papers on improvements in undergraduate science education for non-specialists. Of course, a common method for spreading new ideas about science teaching is the use of innovative textbooks.

In spite of these efforts, it appears many science instructors--perhaps the majority--work in isolation; they do not locate the person or information about teaching improvements that fits their needs and situation. Many others appear to be satisfied with the status quo, perhaps not realizing how much could be done to make science more exciting and more responsive to the educational needs of non-specialists.

Undergraduate science instruction for the non-specialist must be revitalized, and to do so effectively science educators must have information about what others have accomplished.

State academies of science could play an important role at the local level in accomplishing this goal by sponsoring workshops featuring leaders in science education. Such conferences would encourage discussion of teaching ideas and valuable personal contacts. This in turn could lead to follow-up discussions between interested colleagues. These workshops should also involve other scientists, such as industrial chemists, who place a high value on the education of the non-specialist but who are not themselves involved in formal education.

National scientific societies should make sure that, where special teaching journals are lacking, a portion of existing journals be devoted on a regular basis to exchange of ideas about undergraduate science instruction. We recommend that popular science magazines, such as Science 81, Science, and Scientific American, reserve a few pages in each issue for ideas about teaching. Reaction might even be solicited from lawyers, journalists, legislators, and others in the form of special articles or letters to the editor to introduce some feedback into this media forum.

Our Committee has been impressed by the success of
the National Science Foundation's Chautauqua program in
bringing ideas for teaching college science to regional
communities. Supported by the National Science Foundation and coordinated by the American Association for
the Advancement of Science, the University of Georgia,
and 12 regional field centers, Chautauqua forums are
held throughout the United States. Scholars from various fields meet with undergraduate college teachers for
two intensive two-day sessions, typically occurring in
the fall and early spring, with an intervening period of
several weeks for individuals to work on projects related to the course. The primary aim is to enable undergraduate instructors to keep up to date in science
and to expand the relevance of their teaching to today's
world. The program announcement for 1980-1981 reveals
an interesting breadth of lecture topics including "Science, Media, and the Public," "Food, Energy, and Society," "The Changing American Family," "Cognition and
Teaching," and "How Life Began on Earth." In fiscal
year 1981, program support amounted to approximately
$200,000, down from a total of about $1 million in the
previous fiscal year. In the next chapter, we will discuss the important role that the federal government can
play in keeping this program available to the teaching
community.

We discussed above the role of scientific societies
addressing the problems of undergraduate science education. Indeed, some of them have performed this function
for a long time. A review of the role of scientific
societies in the improvement of science instruction for
non-specialists would be incomplete, however, without
mention of the part played by the college science commissions of the 1960s. For a considerable time these
commissions bridged the gap between the individual
science instructor and the rest of the science education
community. With modest funding from the National Science Foundation--about $175,000 per field per year--com-
missions were established in the late 1950s and 1960s in
eight fields: biology, chemistry, physics, geological
sciences, agricultural sciences, engineering, mathematics, and geography. Educators in a particular discipline, including innovative teachers and eminent researchers, were elected to each respective commission
and met approximately four times a year.

Although the commissions' agendas varied, most focused on assisting science faculty members to improve

the teaching of science in two- and four-year colleges and universities. Funds were used to hire core staff, to hold meetings and conferences, and to promulgate ideas for improvement of teaching through newsletters and the like. Attention was divided between the education of future specialists and courses for non-specialists, with the former receiving the lion's share.

These commissions were not involved directly in curriculum development. Instead, they played an important role in spearheading national interest in college science education within their professional communities.

In the early 1970s funding for the commissions ceased due to changing federal priorities and a decreasing federal interest in support of science education (National Research Council, 1981b). This was coupled with a belief that after 5 to 10 years of federal support, the science professions should be ready to pick up the momentum and support the effort. In a few cases, the Committee learned, the activities of the commissions were indeed taken up by the scientific societies; but in most cases the termination of a commission signaled an end to the field's involvement with college curriculum reform. Consequently, initiative was lost.

Participants at the Committee's December 1980 conference concluded that it was neither desirable nor feasible to revive the commissions as they once were. Conferees did agree, however, that in certain fields where there is no central forum to steer national consideration of educational issues, a commission-type mechanism should be considered as a means to initiate discussion.

If the scientific community is to intensify its efforts to improve undergraduate science education for non-specialists, it is important that some entity similar to the college science commissions be in place in each field to provide a mechanism for communication among interested parties. We will discuss why the federal government should play a part in this effort in the next chapter.

5

THE FEDERAL RESPONSIBILITY TO SERVE AS A CATALYST
IN IMPROVING SCIENCE EDUCATION
FOR THE NATION'S FUTURE LEADERSHIP

The question of what role the federal government should
play in American affairs predates the formation of the
Union itself. In the months preceding and following
adoption of the Constitution, politicians and pamphle-
teers waged a fierce debate over how involved the central
government should be in such matters as finance, com-
merce, and military protection. From the very outset of
the controversy, when such stalwarts as Richard Henry
Lee and Alexander Hamilton argued over states' rights,
down to the current debates over Ronald Reagan's new
federalism, there has been almost unanimous agreement
over one point: When it is clearly in the national
interest, the federal government should take decisive
steps to solve problems that plague the republic as a
whole.

In the present case the Committee is convinced that
we are confronted by an educational problem of national
significance and that federal action is warranted. Our
study indicates that, in general, the nation's colleges
and universities are not doing enough to prepare our
future civic and professional leaders with the understanding and knowledge of science that they will need in
order to function effectively. In a sense this is
ironic. Our educational system has graduated experts
who have created a scientific and technological milieu
so complex that other intelligent graduates of these
very same institutions are incapable of comprehending
it. In essence, then, we have reached a point where
even so adamant an anti-Federalist as Thomas Jefferson
would call for federal action. After all, it was he who
wrote to Dr. Benjamin Rush: "I have sworn upon the
altar of God, eternal hostility against every form of
tyranny over the mind of man" (Jefferson, 1800). Our

study reveals an educational deficiency that contributes to an imminent danger of reaching that tyrannical state.

If it makes sense for the federal government to spend billions of dollars in creating one of the most extensive and powerful military-technological establishments in the history of mankind, it also makes sense to dedicate relatively few dollars in an effort to help educate citizens so they can make intelligent decisions about what President Eisenhower called our "overwhelming military-industrial complex" (Eisenhower, 1961).

The time has come for the federal government to take action to help correct this situation. We are not suggesting massive federal intervention nor action solely by the federal government. That would neither prove the panacea some might think, nor would it be in keeping with our belief that education should be the business of educators.

What we are suggesting is a reasonably restrained role in which the federal government would assume a catalytic function and stimulate action. The federal government should also help coordinate efforts across the 50 states and serve as a central clearinghouse for exchanging information and ideas as to how we can best solve our problem. Most important of all is the clarion function. Leaders in the federal government--as high up the prestige scale as the White House itself--need to point up our growing science illiteracy problem and call for concerted action to rectify this educational problem.

In short, we believe the federal government should avoid taking on roles that the states, the private sector, the educational establishment, or individuals can do for themselves. It should, however, in our opinion assume a central catalytic role to make sure that the problem of science education for the non-specialist is addressed on a national scale. In keeping with this philosophical scope, the following recommendations are offered in the belief that they can be carried out with modest funding and appropriate jurisdictional authority.

> The federal policy of program coordination
> and support is needed to strengthen the
> college education of non-specialists in
> science and technology.

Our primary and overriding recommendation stems from a
substudy the Committee launched during its deliberations
to determine just who is doing what at the federal level
to support science education for the non-specialist.
The picture that emerged from our investigation is
this: The federal government is engaged in a diverse
set of science education activities for non-specialists,
but these endeavors are variously directed and lack
coordination.

We believe it would be in the best national interest
to consolidate these activities to improve undergraduate
science instruction in a more efficient manner. We
think there is some role for the National Aeronautics
and Space Administration (NASA) and the Department of
Energy, to name but two agencies, to join with the National Science Foundation in providing a strong program
of support in the area of undergraduate science instruction for non-specialists. Before this can be accomplished, it will be necessary to establish a policy of
federal support for this activity and to name a single
agency to take the lead in these efforts. In short, a
federal commitment will be necessary to effect the
change that is needed in the current haphazard pattern
of federal support.

RECOMMENDATION 7

The Committee recommends that the federal government
focus its efforts to oversee the improvement of undergraduate education for non-specialists in science and
technology by establishing a vigorous program in the
National Science Foundation for this purpose. The Foundation should also be given responsibility for establishing a clearinghouse and for monitoring the diverse
activities of the various federal agencies that are
operating in this area. Most important of all, we urge
the Foundation to assume this leadership role with con-

<u>siderably more dedication and aggressiveness than it has</u>
<u>heretofore displayed toward advancing science education</u>
<u>for non-majors.</u>

This recommendation grows out of our finding that a
surprising number of federal agencies engage in activities that bear directly or indirectly on the quality of
undergraduate science instruction for non-specialists.
In fact, our survey shows that a total of at least $133
million was allocated for programs that had some bearing
on the education of non-specialists in the present fiscal year (see Appendix C), although only $10-15 million
directly impinged on the needs of the non-specialists.
These include numerous programs of the Science and Engineering Education Directorate of the National Science
Foundation; the Fund for the Improvement of Post-Secondary Education and the National Institute of Education, both of which are located in the Department of
Education; and the National Endowment for the Humanities. More to the point, however, none appear to be
engaged in these activities to deliver identifiable
program support for college science education of non-
specialists. In virtually every instance, the support
is an ancillary activity, an extension of an agency's
concern with a broader population or a more general educational function. Thus, total federal support for
direct amelioration of the situation discussed in this
report is undoubtedly a small fraction of the amount
cited above.

We have also identified a number of other federal
agencies whose educational activities, while fragmented,
occasionally bear on undergraduate science instruction
for non-specialists. For example, the Environmental
Education Act of 1970 (P. L. 91-516) and the National
Environmental Policy Act of 1969 (P. L. 91-190) have led
the Department of the Interior to generate course materials for use primarily by secondary school teachers but
also by post-secondary science instructors. Some of
these materials are used in introductory-level college
science courses involving science and non-science majors
alike. Similar educational materials are produced by
the Department of Agriculture, the Department of Com-
merce, and the National Institutes of Health.

While some federal agencies have engaged in the development of educational materials as a result of federal mandate, several other agencies have contributed to
the improvement of college science education for non-

specialists through public affairs activities. NASA and the Department of Energy, through their public or consumer affairs divisions, have established offices of education that engage in a variety of activities all devoted to making more information about science and technology available to the public. NASA, which became heavily involved in science education under the directorship of agency head James Webb in 1960, continues to provide "curriculum support" information designed to serve as resource materials for science instructors. In addition NASA has contracted with Oklahoma State University to provide a traveling program of lectures on space science for all levels of education, including colleges and universities. Similarly, the Department of Energy has extended and broadened the science education activities begun under Atomic Energy Commission chairman Glenn Seaborg in 1961 to include curricular development activities in undergraduate science instruction, although primary emphasis is on kindergarten through high school. To our knowledge, however, there is little communication among these agencies concerning their educational activities in general and none in the area of educating the non-specialist undergraduate.

As can be seen from this brief summary, many agencies are working in isolation from one another without any meaningful communication or coordination. The very fact that we were forced to conduct our own survey of the situation indicates that no one in government is tending the educational store enough to know what is going on elsewhere in government. Such uncoordinated effort can easily result in unnecessary duplication and waste or to sizable gaps in treatment--*hence our recommendation to center and fund coordination of these activities in the National Science Foundation.*

The new NSF program of support for college science education of non-specialists needs to be structured with care.

Statutory authority for the National Science Foundation programs in science education grew out of a concern for an adequate supply of technical personnel rather than out of a desire to support science education per se. NSF policies in science education over the past 20 years have emphasized the education and recruitment of individuals for careers as scientists and engineers, predicated on a belief that "creative science and vigorous, effective technology depend on highly trained, highly talented" individuals (U.S. House of Representatives, 1965).

On many occasions, NSF has construed its definition of "education in the sciences" more broadly to include the education of non-scientists and the general public. However, a declared policy of assistance to colleges and universities in educating non-scientists in science has been lacking. Budgetary evidence for a commitment to this form of science education is also weak. We estimate, for example, that in fiscal year 1979 less than $2 million of the $80 million appropriated for science education activities (or about 2.5 percent) represented projects directed wholly or in part to the improvement of undergraduate science education of non-specialists. We believe that too many years have passed without sufficient attention given by NSF to the education of undergraduate non-specialists in science and technology.

We have already identified a number of ways in which the federal government can assist colleges and universities in carrying out their functions of educating undergraduate non-specialists in science and technology, and we need now to review these.

We believe such a program of support should be built carefully and with consideration around the proposed new NSF program office of undergraduate science education for non-specialists with special attention being given to five goals:

1. The appointment of a staff familiar with the educational needs of undergraduate non-specialists

2. Clearly articulated program goals that are pursued with vigor and persistence

3. Systematic evaluation activities that assess project activities in light of program goals

4. Coordination of information and activities with other federal agencies

5. Sufficient funds for project support and for
carrying out important administrative and monitoring
activities.

The first of these goals is to establish administra-
tively a program unit with a staff familiar with the
issues involved in educating non-scientists. We believe
that consultants and an advisory committee should also
be involved in the design of the program, especially in
its early phases.

Second, the program goals should be clearly articulated, and the types of projects that work toward
achieving those goals should be specifically delineated. We found in the course of conducting our retro-
spective analysis that federal support for the college
science education of the non-specialist was often an
afterthought--an addendum to a program of more general
support often having quite diversified and sometimes
incompatible goals. Unless goals of a program of support are clearly understood, the program may be doomed
to mediocrity or possibly failure.

Third, a program of federal support for the college
science education of non-specialists should have an
evaluation function built into its activities from the
beginning. Regular checks should be made on the feasi-
bility of the program goals in light of performance.
Routine evaluations of projects should also be used to
determine whether projects are meeting the specific
objectives laid out by the program plan. Some consideration should also be given to measuring the impact of
the program on the quality of post-secondary science
education for non-specialists. If one of the purposes
of the program is to support innovative projects that
can be taken up by others, an assessment should be made
of the extent to which that goal is being met.

We also hope that a federal program of this type
would consider as one of its purposes the coordination
of information about support for post-secondary science
education for non-specialists being provided by other
federal agencies (such as the National Endowment for the
Humanities or the Department of Education) and by the
private sector. This coordination activity could be
either designated an on-going function, or carried out
through annual meetings devoted to this activity, or
both.

Finally, resources should be made available to carry
out important administrative and monitoring activities,

as described above, and for a program of awards of sufficient magnitude to catalyze the change that we believe
is needed in the undergraduate science education of non-
specialists. It is beyond the scope of this Committee
to designate the level of funding required. *It seems
quite clear to us that the proposed level of fiscal year
1982 spending for all NSF science education programs--
less than one-third that in fiscal year 1981--is too low
and should be raised. Within the science-education budget, the estimated 2.5 percent of that budget devoted to
non-specialist science education is too small a fraction. We recommend something in the range of 5 to 10
percent as more reasonable. More important than the
total number of dollars allocated, however, is the
degree of commitment of the federal government to the
goals and the vigor and skill with which the program
staff carry out their catalytic role.* If the program is
successful, the multiplying effect of a modest federal
investment will be expansive.

Faculty development should be given high
priority in federal program support.

---

*We recommend that federal financial support should be
given to faculty development.* By this, we mean a
program of support emphasizing at least two components:
an incentive function and an information function.
Incentives for excellence in undergraduate science
instruction were suggested in the previous chapter, but
excellence must also be assessed. *To be more specific,
we believe that the federal government should fund
grants of $20,000 to $25,000 each to establish model
programs in a variety of college settings to explore
innovative approaches in identifying, evaluating, and
rewarding college science instructors.* A great deal has
been said over the years about the desirability of evaluating teaching, but very little concrete activity ever
goes beyond shoptalk. Research productivity is so much
easier to quantify in reaching tenure and promotion
decisions. What can be done?
With federal funding, selected colleges and universities might transcend the discussion stage and establish

model programs for (1) the development of teaching-evaluation instruments to measure the judgments of students, alumni, and peers; (2) the development of innovative approaches to integrating teaching assessments in tenure and promotion decisions; (3) the extension of self-evaluation techniques for science education; and (4) the use of workshops and in-service training to improve teaching.

Innovative assessment procedures should go beyond the use of student-based evaluations. Alumni, who have the advantage of distance and maturity, could provide valuable insights into the teaching contributions of science faculty members. Through campus alumni offices or placement offices, graduates from non-science fields could be surveyed on a regular basis to determine the extent to which courses and faculty contributed to their understanding and use of science and technology. It also is conceivable that properly designed surveys of employers of graduates could be used when decisions of tenure and promotion are made in science departments. A model program could also explore the use of peer evaluation in identifying teaching excellence.

We believe that many individuals have the potential to be excellent teachers but simply lack the opportunity to perfect those skills. We would view an important element in any model program to be the exploration of techniques designed to enhance teaching abilities. Many institutions have begun to use videotaping as one approach to self-improvement and self-evaluation. Linked to a larger program involving in-service training, workshops, and the use of educational consultants, these self-evaluation tools suggest that undergraduate science instruction of non-specialists and specialists alike could be vastly improved.

We also believe that an important dimension would be added to a national commitment to excellence in college science teaching if the White House Award were to reward outstanding classroom performance. *Therefore, we call upon the president to establish and give national recognition to an annual White House Award of at least $5,000 to a teacher who has been selected on a national basis for doing a superior job of teaching science to non-specialists.* Likewise, we urge each of the 50 states to grant Governors' Awards of $2,000 to $5,000 for similar service and achievement as a feeder apparatus into the federal award system.

Another important aspect of faculty development is the opportunity to congregate with colleagues in seminar settings to hear about new teaching ideas from national experts and to exchange thoughts about scholarship and teaching with peer instructors. As mentioned earlier, the Committee has been impressed by the Chautauqua series as a means of stimulating teaching ideas and disseminating innovations for the science classroom. *On the basis of this quality performance record, we recommend that support for the Chautauqua series be restored to about $1 million per year by the National Science Foundation.*

We also believe that the systematic dissemination of information about existing undergraduate science education courses and approaches can play an important role in faculty development. A regularly updated national directory of teaching innovations in college science education for non-specialists would be useful to teachers as a starting point in finding out what other faculty are doing in their courses. Such a directory would permit educators to gain information about the scope of teaching developments at any particular time. Properly compiled, such a directory might offer interesting summaries of programs being undertaken at various institutions across the nation, the goals of these programs for the non-specialist, the materials being used or developed, the texts adopted, the types of laboratories and demonstrations being developed, and evaluations of their performance.

We suggest that the federal government seek out an organization through the National Science Foundation to establish such a directory and fund it at an appropriate level to create a quality communication link among the nation's teachers of science for non-majors. The federal government has supported such efforts in the past. These include a summary of programs and courses on the subject of the ethics and values of science and technology, compiled by the American Association for the Advancement of Science in the late 1970s, and the international directory of science and mathematics curriculum projects maintained by David Lockard at the University of Maryland in the 1960s and 1970s. Clearly, then, a focused national directory of programs for the non-specialist is feasible.

> The federal government should give moral and
> financial support to worthy experimental
> efforts to develop new courses for science
> non-majors.

In Chapter 4, we suggested that college science faculty
should redirect their attention to the development of
effective introductory science courses for undergraduate
non-specialists and to the updating of special topics
courses. We believe the federal government can play an
important part in identifying innovators and assisting
them in such curriculum development.

In the 1960s and 1970s the federal government and
numerous private and industrial foundations directed
funds to the improvement of science education at all
levels. These funds permitted innovators to have at
their disposal the resources necessary to develop curricula and materials for the advancement of science
teaching. A substantial portion of the funding was
directed to the improvement of science education in our
nation's secondary schools. However, some funds were
directed to college science education.

The proportion of awards devoted to the improvement
of college science education of non-specialists was
never great. Having reviewed course improvement projects supported by the National Science Foundation during
that time, we estimate that awards directed specifically
to the improvement of science education for undergraduate non-specialists never exceeded 15 percent of the
total number of projects supported in any one year
(National Research Council, 1981b).

A part of the charge to the Committee was to determine the extent to which any formal efforts of the past
two decades to improve undergraduate science instruction
for non-specialists linger today. Such information
could play an important part in determining new directions for funding support.

To carry out this assessment, the Committee conducted
a series of interviews with 10 former project directors
to determine their views on the success of the projects
(National Research Council, 1981b). The Committee
asked, "To what extent have the results of those projects remained a part of the undergraduate curriculum,

and to what extent have the results been taken up by others?" Projects included in the analysis were restricted to those funded some time between 1960 and 1975 by public or private sources. The primary focus of the projects was the undergraduate non-specialist.

The ten projects that were reviewed (and the sources of support) were "Chemistry for Those Who Would Rather Not" (Lilly Foundation), "Humanistic Approach to the Natural Sciences" (NSF), "Nature of Evidence" (Exxon Education Foundation), "Introductory Physics Sequence" (NSF), "History of Physics Laboratory" (NSF, Sloan Foundation), "Physics of Technology Modules" (NSF), "Geography in Liberal Education" (NSF), "Science Courses for Baccalaureate Education" (Kettering Foundation), "Core Program in Biology" (NSF), and "Physical Science for Non-Science Students" (NSF).

We concluded that, with a few important exceptions, large-scale curriculum improvement projects have not been successful in spreading to institutions other than the ones in which they were developed. This appears to be due to at least two factors. The first is that experiments in curriculum improvement have not succeeded where institutional commitment to curriculum improvement is lacking. For example, only 7 of the 18 colleges originally involved in the project continue to use the "helical course" approach of the "Introductory Physics Sequence" developed by Donald DeGraaf at the University of Michigan in Flint. This physics sequence is a four-semester course designed so that a student can enter at any level of the sequence, depending on prior preparation. The first two semesters are tailored for the non-science major, while those with prior physics experience can step into upper levels of the sequence. DeGraaf concluded that the "commitment of the physics department" is necessary if this sequence is to work in other colleges (National Research Council, 1981b).

Another element that appears to contribute to the failure of innovative approaches to catch on is the lack of sufficient support to permit follow-up activities in colleges interested in trying out new teaching innovations. Follow-up activities are important to answer the questions raised by college faculty experimenting for the first time with these new approaches or to show college faculty how a portion of a course is intended to work.

V. L. Parsegian described his experience in assisting college faculty to use the "Science for Baccalaureate

Education" course he developed at Rensselaer Polytechnic Institute in the late 1960s. This course sought to interrelate the biological and physical sciences for non-science majors within the common theme of thermo-dynamics. A textbook, a laboratory guide, and a teacher's manual were produced. According to Parsegian, this is a very difficult course to teach because of its broad conceptual framework. Faculty are almost required to abandon their disciplinary orientation in favor of more "philosophical conjecture" (National Research Council, 1981b). Parsegian believes that "specific emphasis on teacher training" would have helped faculty adopt this non-traditional approach to undergraduate education. It continues to be used in a modified form by only a few of the original colleges participating in the experimental period.

Curriculum projects that attempt to foster non-tradi-tional thematic approaches to undergraduate science edu-cation for non-specialists seem particularly in need of in-service follow-up support. James V. Connor stressed the potential role for teacher training in assisting faculty to incorporate his "Humanistic Approach to the Natural Sciences." When he began this experiment at a small liberal arts college in the late 1960s, its goal was to develop an interdisciplinary course for students to fulfill their general education requirements. The course emphasized the relationship between science and non-science fields. The main thrust was to "motivate students to see how important science is in their own areas" (National Research Council, 1981b). Connor, who continues to respond to inquiries about the course, be-lieves that weekend workshops and other forms of teacher training would go a long way to assist college faculty in experimenting with this course.

Some innovative approaches to undergraduate science instruction for non-specialists may be doomed to failure because there simply is no market for them. For ex-ample, it is possible that curriculum projects that represent historical or interdisciplinary experiments for teaching science suffer from the same problems that prevented James Conant's historical case-study approach at Harvard from catching on. Students whose only experience with science is through twentieth-century technology cannot identify with the comparatively primi-tive conditions that led earlier scientists to develop de novo the principles forming the basis of modern physics, chemistry, or biology (Doty and Zinberg,

1973). There is also the possibility that inter-disci-
plinary courses have not become more widespread because
the teachers of such courses have themselves been too
narrowly trained. And, of course, there is always the
possibility that some projects were simply not good.
The limited number of cases of this type of approach in
our sample did not permit us to explore these possible
barriers, although the topics merit more research.

Evidence from our study of past curriculum projects
indicates that a type of project that appears to have
wide appeal is one that aims to develop "modules" for
science instruction. These modules are usually packages
for instruction that include a background text, laboratory exercises, learning objectives for the students,
and materials for testing students' mastery of the subject.

Philip DiLavore, together with a number of colleagues, received support from the National Science
Foundation in the early 1970s to develop "modular materials for an introductory non-calculus physics course."
The "Physics of Technology Modules" are built around
familiar devices--a toaster, an ignition system, a loud-
speaker, a fluorescent lamp. They are designed to promote "hands-on" experience so that students can learn by
doing. According to DiLavore, over a five-year period
200-300 colleges and universities and numerous high
schools have used the modules in some fashion (National
Research Council, 1981b).

The primary attraction of "modules" as far as the
teacher is concerned appears to be the freedom to pick
and choose the materials for the class. The faculty
member may wish to adopt the course entirely or merely
to supplement a traditional course with a limited number
of modules.

*We recommend the federal government fund projects to
explore ways to develop substantive courses for non-
specialists having little prior experience with basic
science. Such courses should emphasize firsthand experience with phenomena, laboratory exercises, and demonstrations that are relevant to the needs and experiences
of non-science majors.* Some consideration could even be
given to converting existing high-quality high school
science curricula--which were designed as first courses
--for use in the college classroom by those who have had
little prior experience with science in high school.

*In a program to support the development of course
content, some portion of the funding should be made*

*available to develop special topics courses that treat
timely issues of importance to the concerns of non-specialists. Grants should not be restricted to science
faculty alone. We believe that many interesting approaches to such topics as the ethical implications of
scientific advances have been developed--and have the
potential of being further developed--by non-science
faculty members, especially when they work in concert
with scientists.* Perhaps a program of grants for the
support of curriculum development could be coordinated
with the National Endowment for the Humanities.

We believe that computer-based courses have the potential to offer an exciting way to teach undergraduate
non-specialists science. We suspect, however, that
little is known about the efficacy of existing courses
or areas of possible need. *We suggest that project
support be made available to evaluate the quality of
existing computer-based undergraduate science courses
with respect to their potential value to non-specialists.* Led perhaps by the National Science Foundation,
together with the Department of Education, federal support for a study of the current status of computers in
science education for non-specialists could serve a
variety of purposes. In an era of rising costs in education, the findings from such a study might result in
the establishment of regional resource centers that
would make computers available to colleges unable to
invest in hardware for their own use or unwilling to
take the plunge without some experimentation.

These are just a few of the areas in which curriculum
development projects might make a difference in the
quality of undergraduate science instruction for non-
specialists.

To benefit science education for non-specialists fully requires a cooperative approach by
educators, states, industry, and foundations
as well as the federal government.

The dominant factor in the equation of making science
education for non-specialists work, of course, is the
academic institution. Obviously, no significant changes

in the quality of undergraduate science instruction can occur in the absence of commitment to change by individual instructors, the science department, the college, and the academic administration. This kind of support provides the recognition for teaching achievements and the resources educators need to realize their teaching goals.

Colleges and universities will need assistance, however, in creating a climate within which science education can flourish. It is in this capacity, then, that the federal government--together with the states, private foundations, business, and industry--can help.

The new federalism of the 1980s may be expected to return to our 50 states powers that until recently had been preempted by certain programs of federal support. States will now be expected to raise revenues and set priorities for program expenditures in keeping with the perceived needs of their own residents. We have been impressed with the sensitivity of the various state commissions on higher education with which we have had dealings over the past year. We believe they will play an important role in developing programs appropriate for educational support within their states. We would hope that as a result of the work of this Committee, greater priority will be placed on the improvement of undergraduate science education for non-specialists. Such a program would include the provision of financial incentives to encourage excellence and innovation in science teaching and to make possible interdisciplinary faculty conferences to explore new directions for undergraduate science curricula, as we suggested in the previous chapter.

Realistically, however, it is not at all clear that the undergraduate science education of non-specialists will emerge as an activity of high priority at a time when state support for many social programs will be tight. *Given the immediate need to upgrade science education in our colleges and universities, we believe the federal government should monitor the results of the new federalism closely in this regard and determine appropriate ways to assist colleges and universities to meet their science educational obligations in the event that state support is not forthcoming.*

We urge private foundations, businesses, and industry to assist colleges and universities to meet their obligations to provide appropriate science education to

undergraduate non-specialists. Kenneth Klivington, program officer with the Alfred P. Sloan Foundation, told
an audience of science educators early in 1981 that the
Foundation is eager to revive its commitment to science
education but has not "identified any attack on those
problems which makes sense for an institution of its
size" (Klivington, 1981). We hope the Sloan Foundation
and others will be able to support innovators, to
encourage and reward excellence in teaching science to
non-specialists, and to foster discussions between scientists and non-scientists about the new directions that
science education needs to take. It is clear to us,
however, that the role of these private sources of funding will always be limited by virtue of the nature of
their private entity. Whereas the federal government is
obligated to serve the needs of the nation, the obligation of many private sources is first and foremost to
the goals of their charters--which are not always conso-
nant with national needs. Furthermore, the support of
private foundations can be capricious, changing from year
to year as emerging needs catch the imagination of boards
and officers. The instability of private support often
prevents many innovators from seeking such funds, and
there is no reason to believe that this situation will
change in the near term. *To the extent that private
mechanisms are flawed or fail, federal support should be
forthcoming, in our view.*

In the final analysis, however, we believe that the
federal government can be most useful by serving as a
catalyst to help inspire and move all of these other
segments of American society to act on behalf of improving science education for the non-specialist. To be
sure, it has not escaped the Committee's attention that
these are difficult times for science education because
of federal budget cuts. The steps we recommend are
modest, however, and are not to be taken by the federal
government alone. *An appropriate first step would be
for the federal government to convene a series of meetings to bring together leading representatives from
higher education, state governments, the foundations,
industry, and federal agencies to devise a course of
action for the 1980s and beyond to improve the teaching
of science to non-majors.*

In keeping with the South Sea parable related at the
beginning of this report, an effort should be made to
search the nation over for the wisest sages that can be

found. Together they will have to sit down and discover how to survive the inundating forces that have been explicated in this report. Somehow they will have to learn how to help the non-specialist undergraduates on this nation's campuses not only to survive, but also to master the challenges of science and technology that confront us in the twentieth century.

APPENDIX

METHODOLOGY FOR THE COMMITTEE'S
SURVEY OF THE COLLEGE SCIENCE CURRICULUM

The Committee conducted a catalog survey of the college
science curriculum in the United States. The purpose of
this analysis was to estimate the commitment of four-
year colleges and universities to the science education
of undergraduate non-specialists. Non-specialists were
defined as those persons in undergraduate degree-granting programs at four-year colleges and universities who
do not major in the natural or physical sciences, mathematics, engineering, or health sciences. This population includes, but is not limited to, journalism,
business, liberal arts, and education majors. Institutional commitment to the non-specialist relates to required basic skill preparation in mathematics, general
education requirements in the natural and mathematical
sciences, and to science course electives provided for
or available to non-specialists by science departments
of physics, chemistry, biology, mathematics, and computer sciences. Commitment was described along dimensions of institutional control, level, enrollment,
course type, and Carnegie Classification. A sample of
1979-80 college and university catalogs was analyzed
accordingly. The general research question was: "What
are the opportunities for the undergraduate non-specialist to gain scientific, technical, and mathematical
knowledge during the course of his or her baccalaureate
studies?"

DEFINITIONS

Course Type

Science courses offered by undergraduate science departments were coded according to the instructional goals

89

and targeted student populations as identified by the
catalog course descriptions. Courses in which the
non-specialist was most likely to be enrolled are marked
below with an asterisk (*). The course categories were
as follows:

A. Traditional Subject Matter Courses are those designed to equip the student with an understanding of the
formal subject matter of science.
  1. Science for the departmental major: Courses de-
     signed by a department primarily for their under-
     graduate majors; generally, these are upper divi-
     sion courses.
  2. Science for the science major: Service courses for
     prospective scientists and engineers, often from
     other departments.
  3. Science for health professionals: Service courses
     for pre-meds, nurses, paramedics, technicians, and
     others.
 *4. Science for both the science and non-science
     major: Introductory courses to science subjects
     offered to both the science and the non-science
     majors, e.g., "General Physics," "Introduction to
     Biology."
 *5. Science for the non-science professional: Subject
     matter courses for education majors, business
     majors, humanities majors or other specific non-
     science groups, e.g., "Chemistry for Elementary
     Teachers," "Physics for Architects," "Mathematics
     for Liberal Arts."
 *6. Science for the "non-scientist": Subject matter
     courses for the general "non-science" audience,
     e.g., "Survey of the Physical Sciences," "The
     Phenomena of Life."

B. Special Subject Matter Courses which attempt to
teach science within an integrated or interdisciplinary
framework using a thematic, historical-overview, social-
impact, or popular-topics approach. Some have no college mathematics requirements.
 *7. Science for both the science and non-science
     majors: Special courses for an unspecified
     audience, e.g., "Natural Sciences and the Informed
     Citizen," "Energy, Science and Society."
 *8. Science for the "non-scientist": Special courses
     for non-science majors; audience may or may not be
     specified, e.g., "Perspectives on Computers and

91

Society," "Physics for Poets." Further examples
of these course titles may be found in Appendix B.

Institutional Type

Selection of institutions for the sample to be studied
was limited to those included in the Carnegie
Classification System (1976), which divides
post-secondary institutions into eight major categories
as a function of federal support for academic science,
typical level of degree offered, student enrollment, and
a national student selectivity index, as follows:

A. Research Universities I: The 50 leading universities in terms of federal financial support of academic
science in at least two of three years from 1972-73 to
1974-75, which also awarded at least 50 Ph.D.s in
1973-74.
B. Research Universities II: The top 100 leading institutions in terms of federal financial support in at
least two of three years mentioned above, which awarded
at least 50 Ph.D.s in 1973-74 or were among the top 60
institutions in terms of total number of Ph.D.s awarded
during the years 1965-66 to 1974-75.
C. Doctorate-granting Universities I: Awarded at
least 40 Ph.D.s in at least five fields in 1973-74 or
received at least $3 million in total federal support in
1973-74 or 1974-75. Awarded a minimum of 20 Ph.D.s in
five fields, regardless of the amount of federal support
received.
D. Doctorate-granting Universities II: Awarded at
least 20 Ph.D.s in 1973-74 without regard to field or
awarded 10 Ph.D.s in at least three fields.
E. Comprehensive Universities and Colleges I: Offered
liberal arts programs, as well as programs in such areas
as engineering and business administration but lacked
substantial doctoral programs; institutions in this
group offered at least two professional occupational
programs and enrolled at least 2,000 students in 1976.
F. Comprehensive Universities and Colleges II: Offered liberal arts programs and at least one professional or occupational program; this group included
private institutions with less than 1,500 students and
public institutions with less than 1,000 students in
1976.

G. Liberal Arts Colleges I: Ranked high on a national index of student selectivity or were among 200 leading baccalaureate-granting institutions in terms of the number of the graduates receiving Ph.D.s in leading doctorate-granting institutions from 1920 to 1966.

H. Liberal Arts Colleges II: Liberal arts institutions not meeting the criteria for inclusion in the first group of liberal arts colleges.

RESEARCH QUESTIONS

The inquiry was intended to answer the following questions within categories of institutional control (public or private), level (university or four-year college), undergraduate enrollment, and Carnegie classification.

1. What proportion of the non-specialists' total graduation hours is devoted to general education, i.e., course work intended to meet distributive or breadth requirements as contrasted with concentration?

2. What proportion of the non-specialists' total general education hours is devoted to the biological or physical sciences?

3. What proportion of all institutions require general education in the natural sciences?

4. Approximately how many undergraduates are required to take general education in the natural sciences?

5. What proportion of the total science course offerings are available for election by the non-specialist by science field and course code?

6. What proportion of total institutions offer at least one science course to the non-specialist as a function of science field and course code?

METHOD

A stratified random sample (Table A-1) of 215 four-year colleges and universities was drawn selected from the pool of 1,350 Carnegie Classified Institutions. The sample was stratified by control of institution, level of institution, and undergraduate enrollment.

The sample thus includes 12.3 percent of the 1,748 undergraduate baccalaureate-granting institutions in the U.S. reported by the National Center for Education Statistics (NCES) in 1979.

TABLE A-1 Cell Size and Sample Size of Undergraduate Institutions By Type of Control, Level, and Enrollment

| Control and Level | Undergraduate Enrollment Size | | | | | | | | | |
| | 1-2499 | | 2500-4999 | | 5000-9999 | | 10,000+ | | Total | |
	Cell Size	Sample Size	Cell Size	Sample Size	Cell Size	Sample Size	Cell Size	Sample Size	Cell Size	Sample Size
Public										
University	1	1	3	2	17	12	75	30	96	45
Four-Year College	149	12	126	19	119	26	48	24	442	81
Private										
University	4	3	27	10	27	12	7	6	65	31
Four-Year College	1037	32	91	17	15	8	2	1	1145	58
TOTAL	1191	48	24	48	178	58	132	61	1748	215

Source for institutional data: National Center for Education Statistics, 1979.

93

94

In determining the sample size of each cell, a scheme
was applied which assumed that each institution in a
cell had undergraduate enrollment equal to the average
undergraduate enrollment in that stratum. The sample
frequency for each cell varied as the square root of the
estimated enrollment within that cell. The resultant
sampling proportion was then multiplied by the total
sample size (215) to yield the number of institutions
selected in each cell:

E_i = estimated enrollment per cell

P_i = sampling proportion per cell

N_i = sample number per cell

N_T = total sample number (215)

$$P_i = \sqrt{E_i} \, / \, \sum_{i=1}^{s} \sqrt{E_i}$$

$N_i = P_i \times N_T$

To assure geographic representation within each cell,
a distribution of regions was used to determine a geographic quota for sampling. Finally, a table of random
numbers was used to select institutions for inclusion in
the study. To assure comparability among institutions,
only those colleges listed in the Carnegie Council's
1976 Classification of Institutions of Higher Education
were included for analysis (see Table A-2, pp. 96-101,
for a list of institutions included in the survey).

DATA COLLECTION AND ANALYSIS

Catalogs for the academic year 1979-1980 were analyzed
for each institution in the sample and data recorded on
a protocol form. Information collected included pre-
and post-admission requirements in the basic skills of
mathematics, general education requirements, and data
about the science electives system. The electives
system included science courses available to the
non-specialist in physics, chemistry, biology, mathematics, and computer science departments. The total
number of undergraduate science courses, including

multiple-level ones, was also recorded within each science field. Only graduate level courses were excluded from the study. Data were tabulated to yield simple frequencies.

An analysis of the general education requirements was conducted. Over 90 percent of the institutions surveyed were found to have some form of general education or distributive requirement in place, as Table A-3 (p. 102) illustrates. An analysis was also made of the proportion of general education requirements devoted to a study of the natural sciences. The summary statistics of that analysis may be found in Chapter 3.

Tabulations were also made of the science elective system in the various science fields under study. Table A-4 (p. 103) provides an analysis of course distribution by Carnegie type. Summary tables have been provided in Chapter 3.

TABLE A-2   Four-Year Undergraduate Institutions Included
in the Survey*

PUBLIC UNIVERSITIES

Enrollment 1-2499
  University of Alaska, Fairbanks (DOC II)

Enrollment 2500-4999
  Texas Women's University (DOC II)
  University of South Dakota, Main (DOC II)

Enrollment 5000-9999
  Mississippi State University (RES II)
  Utah State University (RES II)
  University of Idaho (DOC I)
  University of Maine, Orono (DOC I)
  University of Montana (DOC I)
  University of Wyoming (DOC I)
  Clemson University (DOC I)
  University of New Hampshire (DOC I)
  University of North Dakota, Main (DOC I)
  University of Rhode Island (DOC I)
  University of Nevada, Reno (DOC II)
  North Dakota State University, Main (DOC II)

Enrollment 10,000 or more
  Texas A&M University, Main (RES I)
  Purdue University, Main (RES I)
  University of Minnesota, Minneapolis (RES I)
  University of Arizona (RES I)
  University of North Carolina, Chapel Hill (RES I)
  University of California, Berkeley (RES I)
  University of Colorado, Boulder (RES I)
  University of Hawaii, Manoa (RES I)
  University of Illinois, Urbana (RES I)
  University of Iowa (RES I)
  University of Maryland, College Park (RES I)
  University of Michigan, Ann Arbor (RES I)
  University of Utah (RES I)
  University of Washington (RES I)
  University of Wisconsin, Madison (RES I)
  University of Arkansas, Main (RES II)

*Carnegie Classification Code is enclosed in parentheses
after the name of each institution.

Table A-2 Continued

University of Virginia, Main (RES II)
Auburn University, Main (RES II)
Rutgers, The State University of New Jersey, New
    Brunswick (RES II)
University of Oregon, Main (RES II)
Florida State University (RES II)
Indiana University, Bloomington (RES II)
University of Nebraska, Lincoln (RES II)
Temple University (RES II)
University of Tennessee, Knoxville (RES II)
University of Delaware (DOC I)
New Mexico State University, Main (DOC I)
Kent State University, Main (DOC I)
University of South Carolina, Main (DOC I)
University of Toledo (DOC I)

PRIVATE UNIVERSITIES

Enrollment 1-2499
  Johns Hopkins University (RES I)
  Yeshiva University (RES I)
  Rice University (DOC I)

Enrollment 2500-4999
  Massachusetts Institute of Technology (RES I)
  University of Chicago (RES I)
  Princeton University (RES I)
  Washington University (RES II)
  Carnegie-Mellon University (RES I)
  Rensselaer Polytechnic Institute (DOC I)
  University of Denver (DOC I)
  Texas Christian University (DOC I)
  University of the Pacific (DOC II)
  University of Tulsa (DOC II)

Enrollment 5000-9999
  Columbia University, Main (RES I)
  Yale University (RES I)
  Northwestern University (RES I)
  Duke University (RES I)
  University of Pennsylvania (RES I)
  Stanford University (RES I)
  Georgetown University (RES II)
  Tulane University of Louisiana (RES II)

Table A-2 Continued

Howard University (RES II)
Marquette University (DOC I)
University of Notre Dame (DOC I)
Adelphi University (DOC II)

Enrollment 10,000 or more
University of Southern California (RES I)
University of Miami (RES I)
Syracuse University, New York (RES II)
Brigham Young University, Main (DOC I)
Boston University (DOC I)
St. John's University (DOC I)

PUBLIC FOUR-YEAR COLLEGES

Enrollment 1-2499
Citadel Military College (COMP I)
Savannah State College (COMP I)
University of Wisconsin, Superior (COMP I)
Sul Ross State University (COMP I)
Lincoln University (COMP I)
Langston University (COMP II)
University of Maine, Farmington (COMP II)
Mary Washington College (COMP II)
Pennsylvania State University, Behrend College
    (COMP II)
Kentucky State University (COMP II)
New Mexico Highlands University (COMP II)
Wayne State College, Nebraska (COMP II)

Enrollment 2500-4999
Morgan State University (COMP I)
East Stroudsburg State College (COMP I)
Chicago State University (COMP I)
Rhode Island College (COMP I)
Slippery Rock State College (COMP I)
Winthrop College (COMP I)
Mississippi Valley State University (COMP I)
Indiana University, South Bend (COMP I)
Saginaw Valley State College (COMP I)
Missouri Southern State College (COMP I)
Cameron University (COMP I)
Western State College of Colorado (COMP I)
University of Guam (COMP I)

Table A-2 Continued

_____

Southern Oregon State University (COMP I)
University of Texas, Dallas (COMP I)
Southern University, New Orleans (COMP II)
Alabama State University (COMP II)
CUNY, Medgar Evers College (COMP II)
University of North Carolina, Wilmington (COMP II)

Enrollment 5000-9999
CUNY, College of Staten Island, St. George (COMP I)
Mankato State University (COMP I)
University of Wisconsin, Eau Claire (COMP I)
Central State University (COMP I)
Eastern Washington University (COMP I)
Murray State University (COMP I)
Western Carolina University (COMP I)
Fitchburg State College (COMP I)
Bloomsburg State College (COMP I)
West Chester State College (COMP I)
SUNY College, Oneonta (COMP I)
Florida International University (COMP I)
Marshall University (COMP I)
Jackson State University (COMP I)
University of Wisconsin, Stevens Point (COMP I)
Oakland University (COMP I)
Moorehead State University (COMP I)
Southeast Missouri State University (COMP I)
Lamar University (COMP I)
University of Arkansas, Little Rock (COMP I)
Metropolitan State College (COMP I)
Humboldt State University (COMP I)
Western Washington University (COMP I)
Weber State College (COMP I)
Tennessee State University (COMP I)
Old Dominion University (COMP I)

Enrollment 10,000 or more
University of South Florida (DOC II)
Memphis State University (DOC II)
Youngstown State University (COMP I)
California State University, Long Beach (COMP I)
CUNY, City College (COMP I)
University of Nebraska, Omaha (COMP I)
University of Texas, Arlington (COMP I)
San Francisco State University (COMP I)

Table A-2 Continued

San Jose State University (COMP I)
Indiana University of Pennsylvania, Main (COMP I)
CUNY, Queen's College (COMP I)
Ferris State College (COMP I)
Central State Michigan University (COMP I)
Montclair State College (COMP I)
Portland State University (COMP I)
University of New Orleans (COMP I)
Western Illinois University (COMP I)
University of the District of Columbia (COMP I)
Eastern Kentucky University (COMP I)
University of Texas, El Paso (COMP I)
Southwest Missouri State University (COMP I)
California State Polytechnic University, Pomona
    (COMP I)
California State University, Sacramento (COMP I)
East Carolina University (COMP I)

PRIVATE FOUR-YEAR COLLEGES

Enrollment 1-2499
Anderson College (COMP I)
David Lipscomb College (COMP I)
Augustana College (COMP I)
Lewis and Clark College (COMP I)
Bishop College (COMP II)
Barry College (COMP II)
Carson-Newman College (LIB I)
Occidental College (LIB I)
Thiel College (LIB I)
Eckerd College (LIB I)
Agnes Scott College (LIB I)
Wabash College (LIB I)
Carleton College (LIB I)
Hendrix College (LIB I)
Colorado College (LIB I)
Harvey Mudd College (LIB I)
Scripps College (LIB I)
Middlebury College (LIB I)
Benedictine College (LIB II)
Northwest Nazarene College (LIB II)
New England College (LIB II)
Westbrook College (LIB II)
Seton Hill College (LIB II)

Table A-2 Continued

Marymount Manhattan College (LIB II)
Caldwell College (LIB II)
Stillman College (LIB II)
Hillsdale College (LIB II)
Wilberforce University (LIB II)
Sterling College (LIB II)
Dillard University (LIB II)
Hawaii Pacific College (LIB II)
Sioux Falls College (LIB II)

Enrollment 2500-4999
University of Puget Sound (COMP I)
Iona College (COMP I)
Wilkes College (COMP I)
College of Saint Thomas (COMP I)
University of Richmond (COMP I)
Bucknell University (COMP I)
Saint Francis College (COMP I)
Samford University (COMP I)
Valparaiso University (COMP I)
Xavier University (COMP I)
Concordia College, Moorhead (COMP I)
University of Scranton (COMP I)
Siena College (COMP II)
Calvin College (COMP II)
Sacred Heart University (COMP II)
Harding University, Main (COMP II)
Smith College (LIB I)

Enrollment 5000-9999
Hofstra University (DOC II)
New York Institute of Technology, Main (COMP I)
University of Hartford (COMP I)
Pace University, New York (COMP I)
LaSalle College (COMP I)
University of New Haven (COMP I)
University of Dayton (COMP I)
International American University, San German
   (COMP II)

Enrollment 10,000 or more
Catholic University of Puerto Rico (COMP I)

TABLE A-3 Proportion of Undergraduate Institutions
Surveyed Having General Education Requirements in Place
by Cell

Control, Level and Enrollment	Total Sampled (n)	Total Having General Education or Distributive Requirements
Public		
University		
1-2499	1	1
2500-4999	12	11
5000-9999	3	3
10,000+	32	27
Four-Year College		
1-2499	2	2
2500-4999	19	19
5000-9999	10	10
10,000+	17	16
Private		
University		
1-2499	12	11
2500-4999	26	26
5000-9999	12	11
10,000+	8	7
Four-Year College		
1-2499	30	30
2500-4999	24	24
5000-9999	6	6
10,000+	1	1
Total	215	205

TABLE A-4 Number of Undergraduate Science Courses for Non-Specialists by Field, Institutional Type, and Course Type**

Course Types by Field

Institutional Type	Physics					Chemistry					Biology					Mathematics					Computing				
	4	5	6	7/8	T*	4	5	6	7/8	T*	4	5	6	7/8	T*	4	5	6	7/8	T*	4	5	6	7/8	T*
Research I	86	25	32	62	1125	79	3	5	17	983	85	10	18	44	1668	123	81	15	10	1649	44	8	5	11	671
II	56	18	26	46	727	33	3	7	11	659	82	8	3	39	1111	75	65	4	21	1016	29	6	1	2	400
Total	142	43	58	108	1852	112	6	12	28	1642	167	18	21	83	2779	198	146	19	31	2665	73	14	6	13	1071
Doctoral I	52	11	19	43	679	41	4	5	17	649	46	7	7	42	845	92	80	9	15	984	29	10	2	6	337
II	51	4	12	10	353	32	1	4	4	346	47	5	4	20	547	69	29	4	5	452	12	2	0	1	123
Total	103	15	31	53	1032	73	5	9	21	995	93	12	11	62	1392	161	109	13	20	1436	41	12	2	7	460
Comp I	320	65	57	164	2671	213	20	52	82	2679	333	44	66	128	4317	513	310	31	61	3624	169	33	5	23	1430
II	42	7	9	17	298	31	5	6	6	379	64	11	8	10	561	103	48	7	7	547	21	4	0	1	158
Total	362	72	66	181	2969	244	25	58	88	3058	397	55	74	138	4878	616	358	38	68	4171	190	37	5	24	1588
Liberal I	25	1	10	10	238	21	0	7	2	216	31	0	2	2	299	43	7	2	4	330	7	0	0	0	22
II	31	1	2	8	158	28	2	3	7	181	48	1	5	5	288	60	16	3	2	258	7	0	0	0	17
Total	56	2	12	18	396	49	2	10	9	397	79	1	7	7	587	103	23	5	6	588	14	0	0	0	39
Grand Total	663	132	167	360	6249	478	38	89	146	6092	736	86	113	290	9636	1078	636	75	125	8860	318	63	13	44	3158

*T = total number of undergraduate science courses offered by the departments studied.
**See pages 90-91 for course types.

103

APPENDIX

SELECTED COURSES FOR NON-SPECIALISTS
DERIVED FROM
THE COMMITTEE'S SURVEY OF COLLEGE SCIENCE CURRICULUM

TABLE B-1   Selected Traditional Subject Matter Courses
Adapted for Non-Specialists, by Field of Science

PHYSICS
Survey of the Physical Sciences
Introduction to Experimental Physics
Concepts of Physics
Perspectives in Physical Science I and II
Basic Physics
Physics Zero
Topics in Physics
Fundamentals of Physics I
Environmental Physics
The Exploration of Physical Phenomena
The Scientific Method
CHEMISTRY
Elementary Chemistry
General Organic and Biological Chemistry
Essentials of Chemistry I and II
Chemistry in Our Time
Introduction to College Chemistry
The Promises and Perils of Modern Chemistry
Modern Chemical Science
BIOLOGY
General Biology I
Biology--Principles and Prospects
The Dynamics of Man
Biological Sciences Survey
Man in the Natural World
Introduction to Human Anatomy
The Phenomenon of Life
Biology and Man
Human Anatomy and Physiology
General Biology
MATHEMATICS
Selected Topics in Mathematics
Survey of Mathematics
Finite Mathematics
Elementary Analysis
Survey of Contemporary Mathematics
Survey of Statistics
Quantitative and Analytical Thinking
COMPUTER SCIENCES
Computer Concepts

Source: National Research Council, Survey of College Science
Curriculum, 1981.

TABLE B-2 Selected Traditional Subject Matter Courses
for Specific Non-Science Majors, by Field of Science
and Field of Major

PHYSICS
Teaching Methods and Material in the Physical Sciences
    (Education)
Physical Science for Education Majors (Education)
General Physics (Architecture)
Physics for Architects (Architecture)
Elementary Physical Science (Education)
Astronomy (Liberal Arts)
Physics for Elementary Teachers (Education)
CHEMISTRY
Teaching Chemistry (Education)
Fundamentals of Chemistry (Education)
Elementary Chemistry (Education)
Chemistry for Secondary School Teachers (Education)
Criminalistic Lab (Forensic Science)
Chemistry for Elementary Teachers (Education)
BIOLOGY
Anatomy and Physiology I and II (Physical Education)
Basic Principles of Biology (Education)
Biology for Elementary Teachers (Education)
Human Anatomy and Physiology (Health and Physical
    Education)
School Health Education for Elementary and Secondary
    Teachers (Education)
Human Biology (Liberal Arts)
The Teaching of Natural Sciences (Education)
MATHEMATICS
Math for Elementary Teachers I and II (Education)
Algebraic Structure of the Number System (Education)
Calculus for Business and Economics (Business/Economics)
Finite Mathematics (Behavioral Sciences)
The Teaching of Secondary School Mathematics (Education)
Quantitative Methods for Economics and Management
    (Business/ Economics)
Math for Business Students (Business)
Math for Liberal Arts and Business I and II (Liberal
    Arts/Business)
COMPUTER SCIENCES
Computer Programming for Business (Business)
Computers and Computer Sciences for Teachers (Education)
Principles of Programming with Business Applications
    (Business)
Computer Applications in Education (Education)

Source: National Research Council, Survey of College Science
Curriculum, 1981.

TABLE B-3   Selected Special Topics Courses, by Field
of Science

PHYSICS
Cultural Physics
Physics for Poets
The Physics of Acoustics and
    Music
Intelligent Life in the
    Universe
The Physics of Energy
Physics in Science Fiction
Environmental Studies
Physics and Society
Energy, Science, and Society
Science for Involvement
The Mysterious Universe
Energy and Man
Physics for Music Lovers
Energy: Its Use, Resources,
    and Environmental Impact
The Scientific Revolution
    and Its Impact on Modern
    Thought

CHEMISTRY
Chemistry and Society
Chemistry for Changing Times
The Natural Sciences and the
    Informed Citizen
Forensic Science
Environmental Chemistry
Man and the Technological Society
The Mystery of Matter
The Scientific World
Chemistry for Today I and II
Topics in Chemistry
Over-the-Counter Drugs
Better Gardening Through Chemistry

BIOLOGY
Ecology and Human Society
The Genetic Future of Man
Current Crises in Human Survival
Drug Use and Abuse
Bioethics
Biology and the Citizen
Biology and Human Values
Concepts in Biology
Scientific Entomology
Food and Drugs
Sex, Reproduction, and Population
Biology in History

MATHEMATICS
Math and Culture
The Nature and Relevance
    of Math
The History of Mathematics
Math and the Environment
Math and the Modern World
The Structure of Mathematics
Mathematics: A Human Endeavor

COMPUTER SCIENCES
Computers in Society
Perspectives on Computers
    and Society
Computers and Modern Society

Source: National Research Council, Survey of College Science
Curriculum, 1981.

APPENDIX

SELECTED FEDERAL PROGRAMS SUPPORTING COURSE CONTENT
DEVELOPMENT FOR NON-SPECIALISTS
FISCAL YEAR 1981

Little is known about the availability of federal support for the improvement of college science education
for non-specialists. In the absence of an available
data base, the Committee conducted its own limited survey of several federal agencies to determine the extent
to which support for course-content improvement was provided in fiscal year 1981. Through telephone interviews, reviews of program announcements, and face-to-
face discussions with agency staffs, the Committee
determined that at least three federal agencies are
presently supporting science programs aimed in part at
the enhancement of undergraduate science education for
non-specialists. These are the National Science Foundation, the Department of Education, and the National
Endowment for the Humanities. A brief description of
their programs of support is provided in the pages that
follow.

NATIONAL SCIENCE FOUNDATION

As might be expected, the National Science Foundation
provides a focal point for science education in the federal government by supporting research and education to
ensure an increased understanding of science at all educational levels and an adequate supply of scientists and
engineers to meet our country's needs (U.S. Government
Manual, 1980). However, NSF does not support a consoli-
dated identifiable program of college science for non-
specialists. Instead, the Foundation--through the Science and Engineering Education Directorate--has applied
its efforts for non-specialists across a variety of
programs including those labeled "science literacy,"

"public understanding of science," and "science for non-scientists." The Directorate described its two major goals for fiscal year 1981: (1) to help all citizens increase their basic understanding of science and its contributions to the quality of life and (2) to ensure a stable flow of the most talented students into careers in the sciences, with particular reference to increasing participation of minorities and women (NSF, 1980a). Three divisions within the Science Education Directorate include the undergraduate non-science student as a target population directly or indirectly.

Division of Science Education Resources

In fiscal year 1981 the Division of Science Education Resources Improvement (SERI) offered two programs-- Undergraduate Instructional Improvement (UII) and Comprehensive Assis- tance to Undergraduate Science Education (CAUSE)--that supported activities related to undergraduate science education for the non-specialist (NSF, 1980a). Within UII, the Local Course Improvement program (LOCI) provided awards to individuals or small groups of science faculty members for relatively short-term projects concentrating on design, preparation, and evaluation of specific new course materials or teaching strategies. Examples of LOCI projects are "Inquiry Role Approach for Teaching Physical Science" at Kearney State College and "Improvement of Biological Science for the Elementary Teacher" at Arizona State University (NSF, 1979). (Estimated total expenditures in fiscal year 1981 for Undergraduate Instructional Improvement: $6.0 million; for LOCI: $2.7 million.)

CAUSE supported a variety of educational activities in fiscal year 1981, including those designed to affect the education of both science and non-science students and to increase participation of minorities, women, or the physically handicapped in science and engineering. Examples of CAUSE projects include "Improvement of Astronomy Courses and Curriculum Through the Development of an Observational Facility" at Marigold College and "Reform of Freshman Biological Science Laboratory Courses" at Elms College (Development and Evaluation Associates, 1979). (Estimated expenditures in fiscal year 1981 for CAUSE: $8.8 million.)

Division of Science Education Development and Research

The Division of Science Education Development and Research (SEDR) attempts to improve science education at all levels and in all age groups. Projects are limited to the natural and social sciences, mathematics, and engineering. SEDR supports research projects designed to generate new knowledge or to synthesize existing knowledge about science education processes and supports development projects designed to produce, test, and disseminate innovative science instruction. In fiscal year 1981 SEDR had two programs: Development in Science Education (DISE) and Research in Science Education (RISE), both of which supported a limited number of activities related to the undergraduate science education of the non-specialist (NSF, 1980).

DISE supported development, testing, and evaluation of innovative instructional materials; design, testing, and evaluation of innovative instructional delivery modes; and identification of technologies that promise to enhance the effectiveness of science education to include experimentation with and improvement of these technologies. Examples of DISE projects include "Societal Issue-Oriented Physics Modules Project" by the American Association of Physics Teachers and "Use of Micro-computers for Learning Science" at the University of Iowa (NSF, 1980c). (Estimated expenditures in fiscal year 1981: $4.7 million)

RISE has supported research aimed at creating and organizing a body of fundamental knowledge in science education emphasizing two categories of research: the evaluation and synthesis of existing research and its implications and the creation of new knowledge, research methods, and non-quantitative techniques in the empirical sense. Examples of RISE projects are "A Study of Science Instructional Programs in Two-Year Colleges" at the Center for Study of Community Colleges and "Scientific Reasoning: Cognitive Processes in Using and Extending Problem-Solving Skills" at the University of Minnesota (NSF, 1980c). (Estimated expenditures in fiscal year 1981: $6.1 million.)

Division of Scientific Personnel Improvement

NSF's Division of Scientific Personnel Improvement (SPI) is designed to ensure that talented graduate students in

the sciences obtain the education necessary to become
first-line scientific researchers, to train or upgrade
the scientific personnel needed to meet identified national needs, to promote graduate training in institutions traditionally serving ethnic minorities, to provide new knowledge and update experiences for science
teachers, to expose scientifically talented high school
and college students to research activities, and to develop and test methods to stimulate participation in
science by women, minorities, and the physically handicapped (NSF, 1980a). In fiscal year 1981 an estimated
$3 million was provided for college faculty develop-
ment. This included support for College Faculty Short
Courses and Science Faculty Professional Development--
both of significance for the improvement of non-specialist education. The latter program was designed to help
experienced, full-time college science teachers involved
primarily in undergraduate science instruction to increase their competence in science (NSF, 1980a).

DEPARTMENT OF EDUCATION

Through the National Institute of Education (NIE), the
Department of Education supports research on cognitive
development through the Teaching and Learning Research
Program. This program is coordinated with the research
awards program of the NSF Division of Science Education
Development and Research. This joint NIE-NSF program
supports projects in which persons working in cognitive
psychology collaborate with persons from one of the natural sciences, technology, or mathematics to study the
learning and teaching of that discipline. These awards
support research on cognitive processes and the structure of knowledge in science and mathematics and con-
ceivably could lead to the development of course materials that match the content of the science curriculum
more closely with the level of readiness of undergraduate non-specialists to receive the materials (National
Research Council, 1981b). (Current level in fiscal year
1981: $34 million.)

In addition to these cognitive-research awards, the
Department of Education is involved to some extent in
the improvement of undergraduate science instruction for
non-specialists through the Fund for the Improvement of
Post-Secondary Education (FIPSE) administered by the
Office for Post-Secondary Education. Grants are made

through this program to colleges and universities, community colleges, consortia, professional associations, and other groups to improve "organized learning." Among the awards made by the Fund in recent years are a number that bear directly on the enhancement of general education in liberal arts colleges--including the role of science education and the development of practical approaches to teaching science to undergraduate non-specialists. The Comprehensive Program, the Fund's major competitive program, awards grants for a wide range of projects that contribute to better learning, that are cost-effective, and that have the potential for far-reaching influence (FIPSE, 1980b). (Allocation in fiscal year 1981: $13.15 million.)

National Project IV, a FIPSE project in 1979 and 1980, identified, documented, and examined several of the most promising existing programs in liberal education in order to identify a common language for discussion of liberal education, to encourage clarification of the outcomes of such education, and to describe the diversity of curricular and instructional forms which enhance learners' education opportunities. Another activity supported by the Fund, the Mina Shaughnessy Scholars Program, made grants to educational practitioners to reflect on and analyze their experiences in post-secondary education, focusing on nationally significant issues that have emerged in the last two decades (FIPSE, 1980a). (Allocation in fiscal year 1981: $250,000.)

NATIONAL ENDOWMENT FOR THE HUMANITIES

The National Endowment for the Humanities (NEH), through its Division of Special Programs, has engaged in a cooperative effort with the National Science Foundation to support the program in Science, Technology, and Human Values. This program is connected with Ethics and Values in Science and Technology and supports research programs involving both humanists and scientists. Examples include a study of value issues in the control of technology (Appropriations in fiscal year 1981: $1.2 million). The Endowment is seeking to develop similar relationships with other federal agencies that support research and education programs in science and technology.

The Endowment has also helped a number of professional and other schools develop courses and programs in the sciences and the humanities through its Division of Education. The Division of Education Programs has supported a number of curriculum-planning programs designed to develop community college and university programs in technology and society, humanistic imagination and creativity, and the teaching of the history of industrial technology. Through its consultant grants program, the Division has assisted a number of professional and other schools in the development of courses and programs in the sciences and the humanities. In fiscal year 1981 a total of $16.8 million was appropriated for the activities of this Division.

We are grateful for the information provided by numerous agency personnel in compiling this survey. In particular, we would like to thank the following individuals for their assistance in securing budget information for fiscal year 1981: Albert Young, National Science Foundation; Stephen Ehrmann and Andrew Zucker, Department of Education; and John Lippincott, National Endowment for the Humanities.

REFERENCES AND BIBLIOGRAPHY

Abelson, Philip H. "America's Vanishing Lead in
    Electronics." Science, vol. 210, p. 1079, 1980.
Adler, Mortimer J., and Milton Mayer. Revolution in
    Education. Chicago: University of Chicago Press, 1958.
Aldridge, Bill G. "Physics in the Open-Door College."
    Physics Today, pp. 46-51, March 1970.
_____. "National Science Foundation's Other Mission."
    Science, vol. 212, p. 9, 1981.
American Association for the Advancement of Science.
    Ethics and Values in Science and Technology
    (EVIST)--Resource Directory. Washington, D. C.:
    AAAS, 1978.
American Bar Association. Law Schools and Professional
    Education. Report and Recommendations of the Special
    Committee for a Study of Legal Education. Chicago:
    ABA, 1980.
American Chemical Society. "Piaget Takes Hold of
    Chemical Education." Chemical and Engineering News,
    vol. 55, pp. 25-26, 1977.
_____. Chemistry for the Public. Report of the ACS
    Education Conference. Washington, D. C.: ACS, 1978.
Arons, Arnold B. "Using the Substance of Science to the
    Purpose of Liberal Learning." Presented at the
    Symposium on Science in Liberal Education, American
    Association for the Advancement of Science, San
    Francisco, January 1980.
Associated Press. "U.S. Report Fears Most Americans Will
    Become Scientific Illiterates." New York Times,
    October 23, 1980.
Association of American Colleges. The Roles of Science
    and Technology in General and Continuing Education.
    Washington, D. C.: AAC, 1980.

_____. "Technological Literacy and the Liberal Arts." *Forum*, vol. III, 1980.

Association of American Geographers. *A Report of the Geography in Liberal Education Project.* Washington, D. C.: AAG, 1965.

Atkinson, Richard C., and Joseph I. Lipson. "Instructional Technologies of the Future." Presented to the 88th Annual Convention of the American Psychological Association, Montreal, September 1980.

Axtell, James. "The Death of the Liberal Arts College." *History of Education Quarterly*, pp. 339-352, Winter 1971.

Becker, Carl L. *Modern Democracy.* New Haven: Yale University Press, 1941.

Beckwith, Miriam M. *Science Education in Two-Year Colleges: Mathematics.* Los Angeles: Center for the Study of Community Colleges and ERIC Clearinghouse for Junior Colleges, August 1980.

Belknap, Robert. *Tradition and Innovation.* New York: Columbia University Press, 1977.

Bell, Daniel. *The Reforming of General Education: The Columbia College Experience in its National Setting.* New York: Columbia University Press, 1966.

Ben-David, Joseph. *The Scientist's Role in Society: A Comparative Study.* Englewood Cliffs, New Jersey: Prentice-Hall, 1971.

Bennett, William. "Science Hits the Newsstand." *Columbia Journalism Review*, January-February 1981.

Blackburn, Robert, Ellen Armstrong, Clifton Conrad, James Didham, and Thomas McKune. *Changing Practices in Undergraduate Education.* A report to the Carnegie Council on Policy Studies in Higher Education. Berkeley, California: Carnegie Council, 1976.

Bork, Alfred. "Interactive Learning." Millikan Lecture, American Association of Physics Teachers, London, Ontario, June 1978.

Borrowman, Merle L. *Teacher Education in America: A Documentary History.* New York: New York Teachers College, Columbia University, 1956.

Boyer, Ernest L., and Arthur Levine. *A Quest for Common Learning.* New York: Carnegie Foundation for the Advancement of Teaching, 1981.

Bronowski, Jacob. *Science and Human Values.* New York: Harper and Row, 1965.

Burton, Ernest DeWitt. *Education in a Democratic World.* Chicago: University of Chicago Press, 1927.

Bush, Vannevar. Science--The Endless Frontier.
Washington, D. C.: National Science Foundation, 1945.

Carnegie Council on Policy Studies in Higher Education.
A Classification of Institutions of Higher Education.
Revised edition. Berkeley, California: Carnegie
Council, 1976.

_____. Three Thousand Futures. San Francisco: Jossey-
Bass, 1980.

Carnegie Foundation for the Advancement of Teaching.
Missions of the College Curriculum: A Contemporary
Review with Suggestions. San Francisco: Jossey-Bass,
1977.

Carpenter, Ted. Calling the Tune: Communication
Technology for Working, Learning, and Living.
Unpublished report. National Manpower Institute,
Washington, D. C., 1980.

Chaffee, John H., and Thomas B. Evans. Fact Sheet--
Barrier Beaches and Islands. Washington, D. C.: U.S.
House of Representatives, April 28, 1981.

Cohen, I. Bernard, and Fletcher G. Watson. General
Education in Science. Cambridge, Massachusetts:
Harvard University Press, 1952.

Commission on the Humanities. The Humanities in American
Life. Berkeley, California: University of California
Press, 1980.

Council for the Understanding of Technology in Human
Affairs. "Technology for the Liberal Arts." A
workshop summary report from the Massachusetts
Institute of Technology, Cambridge, Massachusetts,
June 1980.

Crane, Verner W. Benjamin Franklin and a Rising
People. Boston: Little, Brown and Company, 1954.

Curti, Merle, and Rod Nash. Philanthropy in the Shaping
of American Higher Education. Newark, New Jersey:
Rutgers University Press, 1965.

Daddario, Emilio Q. "Science Policy: Relationships Are
the Key." Daedalus, vol. 103, pp. 135-144, 1974.

D'Amour, Gene. "The Philosopher as Teacher: Teaching
Philosophy by the Guided Design Method." Meta-
philosophy, vol. 8, pp. 78-86, 1977.

_____ and Charles E. Wales. "Improving Problem-Solving
Skills Through a Course in Guided Design."
Engineering Education, vol. 67, pp. 381-384, 1977.

David, Edward E., Jr. "On the Dimensions of the Technology Controversy." Daedalus, vol. 109, pp. 169-178,
1980.

DeMott, Benjamin. "Mind-Expanding Teachers." Psychology Today, pp. 110-119, April 1981.

Development and Evaluation Associates, Inc. An Evaluation of the National Science Foundation Comprehensive Assistance to Undergraduate Science Education Program (CAUSE). Prepared for the Office of Program Integration, Directorate for Science Education, National Science Foundation. Syracuse, New York: 1979.

Doty, Paul, and Dorothy Zinberg. "Science and the Undergraduate." Content and Context. Edited by Carl Kaysen. A report prepared for the Carnegie Commission on Higher Education. New York: McGraw-Hill, 1973.

Drucker, Peter F. "The Coming Changes in Our School Systems." Wall Street Journal, March 3, 1981.

Edwards, Sandra J. Science Education in Two-Year Colleges: Biology. Los Angeles: Center for the Study of Community Colleges and ERIC Clearinghouse for Junior Colleges, August 1980.

Eisenhower, Dwight D. "Farewell Radio and Television Address to the American People, January 17, 1961." Public Papers of the Presidents of the United States, pp. 1035-1040, 1960-1961.

Eliot, Charles W. The Cultivated Man. Boston: Houghton Mifflin, 1915.

Etzioni, Amitai, and Clyde Nunn. "The Policy Appreciation of Science in Contemporary America." Daedalus, vol. 103, pp. 191-206, 1974.

Feinberg, Lawrence. "Publisher Agrees To Donate $150 Million to Public TV." Washington Post, February 27, 1981.

Fey, James T., Donald J. Albers, and John Jewett. Undergraduate Mathematical Sciences in Universities, Four-Year Colleges, and Two-Year Colleges, 1975-76, vol. V. Washington, D. C.: Conference Board of the Mathematical Sciences, 1976.

Finkelstein, Louis (ed). Thirteen Americans: Their Spiritual Autobiographies. New York: Institute for Religious and Social Studies, 1953.

Fiske, Edward B. "Scientists and Humanists Try Some Conversation." New York Times, February 17, 1981.

Flattau, Edward. "Coastal Time Bombs." New Haven Register, June 4, 1978.

Gallese, Liz Roman. "A Little Calculating and a Lot of Terror Equal Math Anxiety." Wall Street Journal, March 13, 1978.

Glass, Bentley, "Science Education--Process or Content?" Science, vol. 171, pp. 851, 1971.

Goldberg, Fred M., and Gene D'Amour. "Integrating Physics and the Philosophy of Science Through Guided Design." American Journal of Physics, vol. 44, pp. 863-868, 1976.

Goodfield, June. Reflections on Science and the Media. Washington, D. C.: American Association for the Advancement of Science, 1981.

Green, Harold. "Law and Genetic Control: Public Policy Questions," in Marc Lappe and Robert Morrison, Ethical and Scientific Issues Posed by Human Uses of Molecular Genetics. Annals of the New York Academy of Sciences, 265 (1976): 173.

Guralnick, Stanley M. Science and the Ante-bellum American College. Philadelphia, Pennsylvania: American Philosophical Society, 1975.

Haight, G. P. "Balancing Chemistry's Priorities." Change Magazine, vol. 8, pp. 4-5, 1976.

Handberg, Roger and James L. McCrae. "Science Education and the Information About Science and Technology: The Two Cultures Emergent." Journal of Research in Science Teaching, vol. 17, pp. 179-183, 1980.

Hannay, N. Bruce, and Robert E. McGinn. "The Anatomy of Modern Technology: Prolegomenon to an Improved Public Policy for the Social Management of Technology." Daedalus, vol. 109, pp. 25-54, 1980.

Hanson, Norwood. Patterns of Discovery. Cambridge, England: Cambridge University Press, 1958.

Harrison, Anna. "Science and Technology in General Education." The Roles of Science and Technology in General and Continuing Education. Washington, D. C.: Association of American Colleges, 1979.

Hayden, Edward P. "The Luddites Were Right." New York Times, November 14, 1980.

Hoopes, Robert. Science in the College Curriculum. A report sponsored by Oakland University and supported by National Science Foundation. Rochester, Michigan: Oakland University, 1963.

Hughes, Everett C., et al. Education for the Professions of Medicine, Law, Theology, and Social Welfare. A report prepared for the Carnegie Commission on Higher Education. New York: McGraw-Hill, 1973.

Isaacson, Lee E. Career Information in Counseling and Teaching. Boston: Allyn and Bacon, 1971.

Jefferson, Thomas. The Thomas Jefferson Papers. Letter to Dr. Benjamin Rush, September 23, 1800. Manuscript Division, Library of Congress, Washington, D. C.

Jencks, Christopher, and David Riesman. The Academic
Revolution. Garden City, New York: Doubleday, 1968.

Kaplan, Martin. "A Widening Gulf Between Science and the
Rest of Us." Washington Star, May 11, 1981.

Kastrinos, William, and Serena Terziotti. "A Survey of
First-Year Biology Courses." The American Biology
Teacher, vol. 42, pp. 44-45, 1980.

Kaysen, Carl (ed). Content and Context. A report
prepared for the Carnegie Commission on Higher
Education. New York: McGraw-Hill, 1973.

Kimche, Lee. "Science Centers: A Potential for
Learning." Science, vol. 199, pp. 270-273, 1978.

Klivington, Kenneth. "How the Sloan Foundation's Science
Education Programs Are Created, Managed and Moni-
tored." Presented to the Annual Meeting of Directors
of NSF Science Education Development and Research
Grants, Washington, D. C., February 6, 1981.

Kormondy, Edward J. "College Faculty Oriented Programs
of the National Science Foundation: Then, Now, and
Next." Paper prepared for the National Science
Foundation, Washington, D. C., July 1979.

_____, William Kastrinos, and Gertrude G. Saunders. "A
Survey of First-Year College Biology Courses." The
American Biology Teacher, vol. 36, pp. 217-220, 1974.

Lauridsen, Kurt V. (ed). New Directions for College
Learning Assistance, Examining the Scope of Learning
Centers. A Quarterly Sourcebook, Number 1. San
Francisco: Jossey-Bass, 1980.

Levine, Arthur, and John Weingart. Reform of Undergraduate Education. San Francisco: Jossey-Bass, 1973.

Lindquist, Jack (ed). Increasing the Impact. Battle
Creek, Michigan: The W. K. Kellogg Foundation, 1979.

Lockard, David (ed). Eighth Report of the International
Clearinghouse on Science and Mathematics Curricular
Developments 1972. College Park, Maryland: Science
Teaching Center, 1972.

_____. A Report on the Exploratory Conference on the
National Science Foundation's Impact on U.S. Science
Curriculum Development. Unpublished manuscript.
University of Maryland, College Park, Maryland, 1974.

_____. The Tenth Report of the International Clearing-
house on Science and Mathematics Curriculum Development 1977. College Park, Maryland: Science Teaching
Center, 1977.

Magarrell, Jack. "Colleges Offered 15 Pct. More Courses
This Year, Survey Finds; Remedial Classes Increase 22
Pct." Chronicle of Higher Education, June 1, 1981.

119

_____. "Universal Access to Personal Computers Is Urged for College Students, Professors." Chronicle of Higher Education, January 19, 1981.

Mallow, Jeffry V. Science Anxiety: Fear of Science and How To Overcome It. New York: Thomond Press, 1981.

Maugham, W. Somerset. The Summing Up. London: William Heineman, Ltd., 1938.

McGrath, Earl J., ed. Science in General Education. Dubuque, Iowa: Wm. C. Brown, 1948.

Miller, Jon D., Robert W. Suchner, and Alan M. Voelker. Citizenship in an Age of Science. New York: Pergamon Press, 1980.

Mooney, William T., Jr. Science Education in Two-Year Colleges: Physics. Los Angeles: Center for the Study of Community Colleges and ERIC Clearinghouse for Junior Colleges, August, 1980a.

_____. Science Education in Two-Year Colleges: Chemistry. Los Angeles: Center for the Study of Community Colleges and ERIC Clearinghouse for Junior Colleges, August, 1980b.

Morison, Elting E. Turmoil and Tradition. Boston: Houghton Mifflin, 1960.

_____. Men, Machines and Modern Times. Cambridge, Massachusetts: MIT Press, 1966.

_____. From Know-how to Nowhere. New York: Basic Books, 1975.

Mosteller, Frederick. "Innovation and Evaluation." Science, vol. 211, pp. 881-886, 1981.

Mumford, Lewis. The Myth of the Machine. New York: Harcourt, Brace and World, 1967.

National Academy of Sciences. "Resolution on Science Education." News Report, July 1981.

National Center for Education Statistics. Digest of Education Statistics 1979-80. Washington, D. C.: U.S. Government Printing Office, 1979a.

_____. Fall Enrollment in Higher Education 1978. Washington, D. C.: U.S. Government Printing Office, 1979b.

_____. Education Directory, Colleges and Universities 1979-80. Washington, D. C.: U.S. Government Printing Office, 1980.

National Endowment for the Humanities. Division of Education Programs--Guidelines. Washington, D. C.: U.S. Government Printing Office, 1980.

National Institute of Education. Women and Mathematics: Research Prospectives for Change. Washington, D. C.: U.S. Government Printing Office, 1977.

_____. Future Directions for Open Learning. Washington,
D. C.: U.S. Government Printing Office, 1979.
_____. Teaching and Learning Research Grants Announce-
ment. Washington, D. C.: Department of Education,
1980.

National Research Council. Science, Engineering and
Humanities Doctorates in the United States, 1979
Profile. Washington, D. C.: National Academy of
Sciences, 1980.
_____. What Students and Faculty Have To Say About
Undergraduate Science Education for Non-Specialists.
Proceedings of a regional hearing held by the
Committee for a Study of the Federal Role in College
Science Education of Non-Specialists, Bloomington,
Indiana, November, 1980. Unpublished report. Commis-
sion on Human Resources, Washington, D. C., 1981a.
_____. Looking Back on Efforts To Improve College Sci-
ence Education for Non-Specialists. Proceedings of a
conference convened by the Committee for a Study of
the Federal Role in College Science Education of Non-
Specialists, December, 1980. Unpublished report. Commission on Human Resources, Washington, D. C., 1981b.
_____. Understanding the Science Knowledge Needs of the
Non-Science Professions. Proceedings of an invitational hearing held by the Committee for a Study of
the Federal Role in College Science Education of Non-
Specialists, March, 1981. Unpublished report. Commission on Human Resources, Washington, D. C., 1981c.

National Science Foundation. A Guide to Undergraduate
Science Course and Laboratory Improvements. Washing-
ton, D. C.: U.S. Government Printing Office, 1979.
_____. Science Education Databook. Washington, D. C.:
National Science Foundation, 1980a.
_____. What are the Needs in Precollege Science,
Mathematics, and Social Science Education? Washing-
ton, D. C.: U.S. Government Printing Office, 1980b.
_____ and Department of Education. Science and
Engineering Education for the 1980's and Beyond.
Washington, D. C.: U.S. Government Printing Office,
1980c.

Nevins, Allan. The State Universities and Democracy.
University of Illinois Press, 1962.

Niebuhr, Reinhold. The Children of Light and The
Children of Darkness. New York: Charles Scribner's
Sons, 1953.

Perlman, David. "Science and the Mass Media." Daedalus,
vol. 103, pp. 207-222, 1974.

President's Commission on the Accident at Three Mile Island. "Report of the Public's Right to Information Task Force." Washington, D. C.: The White House, October 1979.

President's Science Advisory Committee. Education for the Age of Science. Washington, D. C.: The White House, May 24, 1959.

Price, Don K. "Money and Influence: The Links of Science to Public Policy." Daedalus, vol. 103, pp. 97-114, 1974.

Rich, Spencer. "Colleges Majoring in Squeezing the Buck." Washington Post, November 20, 1980.

Ritterbush, Philip C. "The Public Side of Science." Change Magazine, September 1977.

_____. Science Education for Public Awareness and Involvement: Comments on the Federal Role. Unpublished manuscript. National Research Council, Washington, D. C., April 1980a.

_____. Problems and Cultivation--Dimensions of Concern About Limits on the Scope of Science in General Undergraduate Education. Unpublished manuscript. National Research Council, Washington, D. C., September 1980b.

Roark, Anne C. "The Playground of the Museum World." The Chronicle of Higher Education, pp. 8-10, February 13, 1979.

Rockefeller Foundation. Toward the Restoration of the Liberal Arts Curriculum. Working papers of a Rockefeller Foundation Conference, June 1979.

Rosen, Sidney. "Charles W. Eliot's Legacy to Science Education." The Science Teacher, December 1980.

Rossiter, Clinton, and James Lare (eds). The Essential Lippman: A Political Philosophy for Liberal Democracy. New York: Random House, 1963. (Reprinted from Lippmann, Walter. A Preface to Politics. New York and London: Mitchell Kennerley, 1913.)

Rudolph, Frederick. Curriculum: A History of the American Undergraduate Course of Study Since 1636. Carnegie Council on Policy Studies in Higher Education. San Francisco: Jossey-Bass, 1977.

Sagan, Carl. "There's No Hint of the Joys of Science." TV Guide, February 4, 1978.

Sawhill, John C. "The Role of Science in Higher Education." Science, vol. 206, October 19, 1979.

Schulhof, Michael P. "Scientists in Business." New York Times, February 1, 1981.

Schwab, Joseph J. Science, Curriculum, and Liberal
    Education. Chicago: The University of Chicago Press,
    1978.

Shaffer, Richard A. "Personal Computers Are Becoming
    More Useful to Many Investors for Managing Port-
    folios." Wall Street Journal, December 29, 1980.

Shein, Edgar H. Professional Education, Some New Direc-
    tions. The tenth in a series of profiles sponsored by
    the Carnegie Commission on Higher Education. New
    York: McGraw-Hill, 1972.

Slaughter, John B. "The National Science Foundation
    Looks to the Future." Science, vol. 211, pp. 1131-
    1136, 1981.

Snedden, David. Cultural Educations and Common Sense: A
    Study of Some Sociological Foundations of Educations
    Designed To Refine, Increase, and Render More Func-
    tional the Personal Cultures of Men. New York:
    Macmillan, 1931.

Stocking, S. Holly. "Don't Overlook the 'Social' in
    Science Writing Courses." Journalism Educator, vol.
    36, pp. 55-57, 1981.

Stout, David K. "The Impact of Technology on Economic
    Growth in the 1980's." Daedalus, vol. 109, pp. 159-
    168, 1980.

Thomas, Lewis. "The Pre-Med Syndrome." Chronicle of
    Higher Education, December 4, 1978.

Thomas, William A. "A Report from the Workshop on
    Cross-Education of Lawyers and Scientists."
    Jurimetrics Journal, vol. 19, pp. 92-99, 1978.

Todd, Lord. "The State of Chemistry." Chemistry and
    Engineering News, pp. 28-33, October 6, 1980.

U.S. Department of Education, The Fund for the Improve-
    ment of Post-Secondary Education. The Mina Shaugh-
    nessy Scholars Programs. Washington, D. C.: U.S.
    Government Printing Office, 1980a.

_____. Resources for Change--A Guide to Projects 1980-
    1981. Washington, D. C.: U.S. Government Printing
    Office, 1980b.

U.S. Department of Health, Education and Welfare, Office
    of Education. Education for the Professions. Washing-
    ton, D. C.: U.S. Government Printing Office, 1955.

_____. Resources for Change--A Guide to Projects
    1979-80. Washington, D. C.: U.S. Government Printing
    Office, 1979.

U.S. Department of Labor, Bureau of Labor Statistics.
    Handbook of Labor Statistics. Washington, D. C.: U.S.
    Government Printing Office, 1980.

U. S. Government Manual 1980-1981. Washington, D. C.:
    U.S. Government Printing Office, 1980.
U.S. House of Representatives. National Science
    Foundation: A General Review of Its First Fifteen
    Years. Committee on Science and Astronautics, 89th
    Congress, 1st Session, 1965.
Van Doren, Carl. Benjamin Franklin. New York: The
    Viking Press, 1938.
Wallis, Claudia. "Trying to Thwart the Fruit Fly." Time,
    July 27, 1981.
Walsh, John. "Harvard, Science, and the Company of
    Educated Men and Women." Science, vol. 202, pp. 1063-
    1066, 1978.
Ward, F. Champion. The Idea and Practices of General
    Education: An Account of the College of the
    University of Chicago by Present and Former Members
    of the Faculty. Chicago: University of Chicago Press,
    1950.
Waterman, Alan T. "Introduction" in Science--The Endless
    Frontier by Vannevar Bush. Reissued on the tenth
    anniversary of the Foundation. Washington, D. C.:
    National Science Foundation, 1960.
Watson, F. G., and I. B. Cohen. General Education in the
    Sciences. Cambridge, Massachusetts: Harvard Univer-
    sity Press, 1952.
Weinberg, Alvin M. Reflections on Big Science. Cam-
    bridge, Massachusetts: MIT Press, 1967.
Westbury, Ian, and Niel J. Wilkof (eds). Science, Cur-
    riculum, and Liberal Education. Chicago: The Univer-
    sity of Chicago Press, 1978.
Willcox, A. B. "To Know Is Not To Teach." Change
    Magazine, vol. 9, pp. 26-27, 1977.
Wolfle, Dael. The Home of Science, The Role of the Uni-
    versity. The twelfth of a series of profiles spon-
    sored by the Carnegie Commission on Higher Education.
    New York: McGraw-Hill, 1972.
Young, M. F. D. (ed). Knowledge and Control; New Direc-
    tions for the Sociology of Education. London:
    Collier-MacMillan, 1971.
Ziman, John. The Force of Knowledge. Cambridge, England:
    Cambridge University Press, 1976.
_____. "Seeing Through Our Seers." Radio broadcast.
    Reprinted in The Listener, June 24, 1976.

PARTICIPANTS IN THE
COMMITTEE'S HEARINGS AND WORKSHOP

STUDENT AND FACULTY HEARINGS
AT INDIANA UNIVERSITY

November 14-15, 1980
Ernie Pyle Hall, Indiana University, Bloomington, Indiana

Members of the Committee
Richard G. Gray, Chairman, Indiana University
   (Journalism)
H. Richard Crane, Co-chairman, University of Michigan
   (Physics)
Johns Hopkins III, Washington University   (Molecular
   biology)

Study Director
Pamela Ebert-Flattau

Speaker
Robert Scott, Associate Commissioner, Indiana Commission
   for Higher Education

Faculty, Indiana University
Michael Carrico, School of Law
Judith Franz, Department of Physics
Donald Kerr, Department of Mathematics
Julia Lamber, School of Law
Edwin Lambeth, School of Journalism
Alfred Ruesink, Department of Biology
Alex Tanford, School of Law
Donald Winslow, School of Education

126

Alumni, Indiana University
Barbara DeWitz (History)
John DeWitz (History)

Students, Indiana University
Catherine Bonser (Economics/mathematics)
Jennifer Crittenden (Linguistics)
Jan Eveleth (Astrophysics)
Cathy Friedman (Speech pathology)
Kitty Grogan (Elementary education)
Julie Jontz (Audiology)
Karen Kovacik (English/Spanish)
Judith Lawrence (Sociology)
James McConnell (Telecommunications)
Stuart Muir (Comparative literature)
Ann Neugebauer (Speech communication)
Patricia Postel (Recreation)
Teresa Richards (Elementary education)
Debbie Rissing (English)
Jill Sandler (Elementary education)
Holly Stocking (Communications)
Douglas Strommen (Economics)
Jason Young (Political science/psychology)

INVITATIONAL WORKSHOP
ON UNDERGRADUATE SCIENCE INSTRUCTION
FOR NON-SPECIALISTS

December 16, 1980
The Lecture Room, National Academy of Sciences
2101 Constitution Avenue, N.W., Washington, D.C.

Members of the Committee
Richard G. Gray, Indiana University
William G. Aldridge, National Science Teachers
  Association
Donald Bitzer, University of Illinois
H. Richard Crane, University of Michigan
Emilio Daddario, Hedrick and Lane, Washington, D.C.
Lucius P. Gregg, Jr., Bristol-Myers Company, New York,
  New York
Anna Harrison, Mount Holyoke College
William B. Harvey, Boston University
Johns Hopkins III, Washington University
Watson Laetsch, University of California
Estelle Ramey, Georgetown University
Richard L. Turner, University of Colorado
David E. Wiley, Northwestern University

Study Director
Pamela Ebert-Flattau

Workshop Participants
Arnold Arons, Department of Physics, University of
  Washington
Henry A. Bent, Department of Chemistry, North Carolina
  State University
Donald Bushaw, Department of Mathematics, Washington
  State University
Homer Folks, College of Agriculture, University of
  Missouri
Edward A. Friedman, Dean of the College, Stevens
  Institute of Technology
Arthur H. Livermore, American Association for the
  Advancement of Science
William H. Matthews III, American Geological Institute,
  Lamar University
Martin Schein, Department of Biology, West Virginia
  University

128

Arnold Strassenburg, American Association of Physics
Teachers, SUNY at Stony Brook
John G. Truxal, College of Engineering, SUNY at Stony
Brook
Harold Winters, Department of Geography, Michigan State
University
Gail S. Young, Department of Mathematics, Case Western
Reserve University

Invited Observers
American Anthropological Association: Thelma Baker
American Chemical Society: Janet Boese
American Geological Institute: A. G. Unklesbay
American Psychological Association: L. Kaplinksi, Kathy
Lowman
American Sociological Association: Lawrence J. Rhoades
Association of American Colleges: Mark Curtis
Association of American Geographers: Sam Natoli
Department of Education: James Rutherford
Federation of American Societies for Experimental
Biology: Robert W. Krauss
Institute of Medicine: Karl Yordy
Mathematical Association of America: A. B. Willcox
National Academy of Engineering: Randolph W. King
National Research Council: J. F. Blackburn, Catherine
Iino, William Kelly, Samuel McKee, William Spindel,
Russell B. Stevens
National Science Foundation: Alfred Borg, Alphonse
Buccino, Rita Peterson

INVITATIONAL HEARING
ON SCIENCE AND THE PROFESSIONS

March 20, 1981
The Board Room, National Academy of Sciences
2101 Constitution Avenue, N.W., Washington, D.C.

Members of the Committee
Richard G. Gray, Indiana University
William G. Aldridge, National Science Teachers
    Association
Donald Bitzer, University of Illinois
H. Richard Crane, University of Michigan
Emilio Daddario, Hedrick and Lane, Washington, D.C.
Lucius P. Gregg, Jr., Bristol-Myers Company, New York,
    New York
Anna Harrison, Mount Holyoke College
William B. Harvey, Boston University
Johns Hopkins III, Washington University
Gerard Piel, Scientific American
Richard L. Turner, University of Colorado
David E. Wiley, Northwestern University

Study Director
Pamela Ebert-Flattau

Panel on Politics
Thomas Mann, American Political Science Association,
    Washington, D.C.
Judith Sorum, independent consultant, Washington, D.C.
William G. Wells, Head, Public Sector Programs, American
    Association for the Advancement of Science

Panel on Journalism
Christine Harris, Director, Consortium for the
    Advancement of Minorities in Journalism Education,
    Northwestern University
Malcolm Mallette, American Press Institute, Reston,
    Virginia

Panel on Law
Harold P. Green, Fried, Frank, Harris, Shriver, and
    Kampelman, Washington, D.C.
Lee Loevinger, Hogan and Hartson, Washington, D.C.

William A. Thomas, Consultant, American Bar Association
Robert B. Yegge, Professor and Dean Emeritus, University
   of Denver, School of Law

Panel on Business and Industry
Jerrier A. Haddad, Vice President for Technical
   Personnel, IBM Corporation, White Plains, New York
Robert P. Stambaugh, Director, University Relations,
   Union Carbide Corporation, New York, New York

Panel on Religion and Philosophy
The Reverend Michael P. Hamilton, Canon, Washington
   Cathedral, Washington, D.C.
David Smith, Chairman, Department of Religious Studies,
   Indiana University
LeRoy Walters, Center for Bioethics, Kennedy Institute,
   Georgetown University

Panel on Education
Hans Andersen, School of Education, Indiana University
David Lockard, Science Teaching Center, University of
   Maryland
Herbert Striner, Dean, School of Business Adminstration,
   The American University

Invited Observers
Joel Aronson, National Science Foundation
William Kelly, National Research Council